## Take Online Tests

- Take four GREs, GMATs, LSATs, or SATs online from home, school or work.
- Simulate the actual computer-adaptive environment of the GMAT and GRE with thousands of sample questions.
- Take linear SATs and LSATs and get valuable feedback on your performance.

# Save 50%

- Get 50% off the list price of online tests with the purchase of this book—a $10–$25 value.
- Get the best electronic test preparation anywhere for less than the price of a book with CD.

## Maximize Your Scores

- Get instant scores. You can start working on your weak areas immediately.
- Get clear and informative score reports and performance analysis to help you work effectively.
- Get detailed feedback on each answer choice to help you hone your skills—and raise your scores.

## Register Today!

- Go to **http://www.review.com/cracking** and start raising your scores today.

# GMAT Math Workout

BY

JACK SCHIEFFER

RANDOM HOUSE, INC.
NEW YORK

www.randomhouse.com/princetonreview

Princeton Review Publishing, L.L.C.
2315 Broadway
New York, NY 10024
E-mail: comments@review.com


Published in the United States by Random House, Inc., New York, and simultaneously in Canada by Random House of Canada Limited, Toronto.

ISBN 0-679-78373-3

Editor: Jeannette Mallozzi
Production Editor: Maria Dente
Production Coordinator: Jennifer Arias

9 8 7 6 5 4 3

First Edition

# CONTENTS

# 1

# Introduction

## WELCOME

You want to get an MBA and you know that you need a good GMAT score to get into your top-choice business schools. However, your math skills are a bit rusty. You may not have taken many (or even any) math classes in college. These days, you probably use a calculator or computer to balance your checkbook, to crunch numbers at work, or to handle any other calculations that come your way. The result is that you haven't really used your math muscles for several years or more. That's the bad news.

However, there is good news. At one point, you did learn the math tested on the GMAT. None of the concepts is more advanced than high school algebra and geometry. No trigonometry, no calculus, and no multi-variable regression analysis (whatever that is). So you don't need to learn a lot of new stuff, just refresh your memory.

Another piece of good news is that you bought this book (if you're reading this in the bookstore, go over to the register and pay up). In the following chapters, you'll cover the math you need to know for the GMAT and only that. If it's not on the test, it's not in this book. You'll also learn some test-taking strategies specific to the GMAT. This stuff probably won't help you in your first-year statistics course, but it will help you to get there in the first place.

## WHAT IS THE PRINCETON REVIEW?

The Princeton Review (TPR for short) is one of the leaders in helping people prepare for standardized tests such as the GMAT. The company started by offering courses and tutoring for the SAT in 1981. Since then, it has expanded into courses, books, and software for all of the major standardized tests.

TPR's central philosophy is that the tests are beatable. Standardized tests measure more about test-taking skills than they do about fundamental knowledge or intelligence. The company is dedicated to teaching people the test-taking skills they need to perform their best and beat the standardized tests.

Glad to meet you!

## THE GMAT AND BUSINESS SCHOOL ADMISSIONS

You already know that you have to take the GMAT in order to get into business school, but there may be a number of other things you're not so sure about. How important is the GMAT? What's a good score? What are other things that schools consider?

The importance of the GMAT depends on several factors. One is how long you've been out of school. If you graduated a long time ago, say more than five years, then MBA programs will place more weight on your GMAT score than they would if you graduated a year or two ago. That's because they will de-emphasize your college GPA in considering your application and so the GMAT score becomes more important in the mix.

Another factor in the importance of your GMAT score is the particular GMAT score in question. In addition to the overall score (200–800 range), you will receive a math score (0–60 range), a verbal score (0–60 range), and an AWA essay score (1–6 range). Most schools concentrate on the overall score and the math score in their admissions decisions. They look at the overall score because it's a broad measure of your ability (or so ETS has told them). They look at the math score because many MBA courses require significant use of quantitative skills and the schools want to ensure that entering students will have the necessary mathematical ability.

## STRUCTURE OF THE GMAT

The GMAT lasts approximately four hours. The test is administered by computer.

| Section | Questions | Time |
|---|---|---|
| Tutorials | —— | No time limit |
| AWA—Issue | 1 essay | 30 minutes |
| AWA—Argument | 1 essay | 30 minutes |
| Math | 37 multiple-choice | 75 minutes |
| Verbal | 41 multiple-choice | 75 minutes |

The tutorial gives you some time to get comfortable with the use of the computer and the testing software. It will show you step-by-step how to navigate the test, that is, how to choose an answer and proceed to the next question. Although the tutorial is not timed, you shouldn't waste a lot of time here. Take the time you need to familiarize yourself with the workings of the test, but don't get distracted and expend all of your energy on the tutorial. The test is already long enough that fatigue is a factor; you don't want to exacerbate the situation. If you can, you should practice the test using ETS's PowerPrep software or a similar software package. That way, you'll already know how the test works before you walk in the door to the testing center.

## HOW A CAT WORKS

The test is adaptive because it tries to match the difficulty of the test to your performance. The better you do, the harder the test gets.

Each section starts with a question of medium difficulty and assigns you a medium score. When you answer a question correctly, the computer raises your score and gives you a harder question. When you choose a wrong answer, the computer lowers your score and gives you an easier question.

At the beginning of a section, the computer knows very little about you, so each answer, right or wrong, can change your score a lot. By the end of the section, in contrast, the computer already has a very strong idea of your performance, so your answers to the last few questions won't change its mind much at all.

Do you really need to know all this? Well, you don't need to be an expert in computer adaptive testing (CAT), but it's good for you to know a little so that you don't freak out when the test gets harder and harder and harder. In chapter 2, we'll look at some of the implications of the CAT format and how you can best attack the Math section.

## HOW TO USE THIS BOOK

Each chapter in this book reviews one particular area of math or type of problem. The chapters are also arranged sequentially so that later chapters assume familiarity with the material covered in earlier chapters. Spaced throughout each chapter are quizzes that will test your knowledge of the material in each subsection. At the end of each chapter is a set of problems that covers all the material in that chapter. Explanations are provided for all the problems in the quizzes and problem sets.

Ideally, you'll work through the book from front to back, taking the time to review each subject area thoroughly, work all of the practice sets, and check all of the explanations. You should give yourself several weeks (at least!) for this plan, so that you have plenty of time to review the practice sets and learn from your mistakes.

Of course, we don't all live in an ideal world, and you may not have the time to review each subject, work every problem, and read every explanation. If this is your situation, you should skip over the topics that you're familiar with. Try a few of the problems. If you are comfortable with them, move on to something else that gives you more trouble. If you're not comfortable with them, review that topic more thoroughly and do most or all of the problems. You should definitely spend some time with chapters 2, 3, 8, and 10. Those chapters contain a lot of test-taking strategies specific to the GMAT that you are unlikely to have seen previously. You still need at least a couple of weeks for this approach.

What if you only have a few days until the test? Put this book down, call your testing center, and reschedule your session for four to six weeks from now. Seriously. There's no way you're going to learn all of this stuff overnight, and attempting to do so will just stress you out and hurt your scores. Rescheduling will cost you a little money (currently $40), but it will give you the time you need to practice and learn.

## CALCULATION PRACTICE

Although the GMAT tests knowledge of concepts and problem analysis more than calculation skills, you're going to do a significant amount of number-crunching during the test. You want to make sure that your number manipulation skills are sharp so that you won't make careless mistakes or waste precious time struggling with the calculations.

Chances are good that you do most of your calculations with the aid of a calculator or a computer spreadsheet. When was the last time you did long division by hand? Hmm, that's what I thought.

Starting today, do all your day-to-day math by hand: Balance your checkbook, figure out a 15-percent tip, calculate your softball batting average, and so forth. You can check your results with a calculator, but force yourself to exercise those math muscles.

## ANSWER CHOICES

On the GMAT, the answer choices don't have any names, there are just little bubbles in front of each choice. In this book, however, we're going to assign letters to each answer choice: (A), (B), (C), (D), and (E). This will make it easier for us to refer to answer choices as we work through questions.

# 2

# General Test-Taking Tips

The GMAT, like all other standardized tests, follows certain predictable patterns. That's why it's "standardized." The test-writers must follow these patterns so that everybody who takes a GMAT gets tested on the same criteria. Otherwise, their scores wouldn't be comparable and schools couldn't use those scores to evaluate applicants.

By learning the methods and patterns that the test-writers follow, you can use some general test-taking strategies that will help you beat the test. These guidelines apply no matter what specific topic the question covers, so they are useful throughout the test. **Don't underestimate how much these tips can help your score.**

## PACING

Time is your most precious resource on the GMAT. You have only 75 minutes to do the 37 questions in the math section; be sure you use your time wisely. You should keep three pacing goals in mind as you take the test:

1. **Answer every question.** There is a significant penalty to your score if you don't answer all of the questions in each section. You should pace yourself so that you have a chance to get to each question. However, if you find that you can't work every question, you should at least guess an answer for every question left. Save the last two or three minutes of each section to guess on any remaining questions.
2. **Start slowly.** On a CAT, the earlier questions are more important than the later questions. So you should allocate your time accordingly. Go slower and be more careful on the earlier, more important questions. Don't spend five minutes on a question, just try to minimize your chances of making a careless mistake. As you progress through the section and get to less and less important questions, you should gradually pick up some speed. If you make a careless mistake on a question near the end, it will have a very small effect on your final score.
3. **Don't waste time on "impossible" questions.** Almost everyone comes across a handful of questions that are too difficult to do. No matter how much time you spend staring at them, you're not improving your chances of answering them correctly. Cut your losses on these questions. Once you realize that it's an impossible question, take an educated guess (more on this later), and move on. Spend your time on the questions that will get you points.

## GUESSING

Some people feel that guessing is a terrible thing to do. It's like giving up. On the GMAT, guessing is something you're going to have to do at some point. Remember, the questions keep getting harder and harder if you continue to answer them correctly. So you're eventually going to run into some questions you don't know how to work. It's okay to guess on these. In fact, it's better to go ahead quickly and guess than to waste a lot of time and end up taking a guess anyway.

## USING SCRATCH PAPER

Because the questions are presented on a computer screen, you'll have to do all of your work on a separate sheet of scratch paper. You can't write directly on the problem to label diagrams, cross-off answers, and so forth. It's very important that you actually use the scratch paper! You might be tempted to do a lot of the work in your head. Don't! That's a sure-fire way to make careless mistakes.

When you copy diagrams, equations, and other information to your scratch paper, double-check it to make sure it's accurate. You should also write A B C D E for each question so that you can physically cross off answers as you eliminate them. This will help you keep track of where you are as you work the problem.

When you practice questions from this book, you should do your work on a separate sheet of scratch paper. This will help you get used to copying stuff to your scratch paper.

## JOE BLOGGS

Choose a number and write it down here: ____________

Don't read any further until you've chosen your number. You'll need to refer to this number in a few minutes.

When the test-writers at ETS write questions for the GMAT, they need to generate questions that cover the range of difficulty from pretty easy to really tough. However, they're restricted by the topics they're allowed to test. They can't write questions about calculus or other difficult topics in order to get hard questions. Instead, they have to use the same topics, yet somehow they have to make the questions hard. One way the test-writers make a question hard is to put in things that will trick you into choosing the wrong answer.

People tend to think in predictable patterns. For example, look at the number you wrote down earlier. You could have chosen any number, including 0.08, –7.5, $\frac{2}{3}$, $\sqrt{7}$, and similar numbers. However, no one chooses those numbers; almost everyone chooses a whole number such as 1, 2, or 3. In addition, most people choose a whole number from 1 to 10. People choose these numbers because they are "programmed" to think that way.

This type of thinking is called the Joe Bloggs response. Joe Bloggs is the completely predictable person who lives inside each of us. When you look at a math problem, watch your initial reaction to the question, your "five seconds or less" solution. That's the Joe Bloggs answer.

The ETS test-writers use those predictable thinking patterns against you. They write questions in such a way that being Joe Bloggs will lead you to the wrong answer. You should be suspicious of a solution that you come up with too quickly. If you didn't use at least 30 seconds to solve the question, you should double-check your answer. That doesn't mean that the easy answers are ***never*** correct, just that they are ***usually*** traps. Look at this next example:

1. Clare's Boutique is a store that sells silk scarves at an everyday 20% discount from the normal list price. During a Mother's Day sale, all the scarves are sold at an additional 10% discount from everyday price. The sale price is what percent of the normal list price?

   (A) 28%
   (B) 30%
   (C) 70%
   (D) 72%
   (E) 80%

The Joe Bloggs instinct says, "Simple. A 20% discount plus a 10% discount is a 30% discount." So Joe would choose (C) or, if he misread the question, (B). So you want to be suspicious of those answers. If you come up with one of those answers, double-check your work and see if there's a different approach you could take. Take a look at another example:

2. If $w + 2x = 150$, $2w + 3y = 100$, and $x + 3z = 50$, what is the value of $w + x + y + z$?

   (A) 12.5
   (B) 20
   (C) 50
   (D) 100
   (E) It cannot be determined from the information provided.

Your first instinct is probably something like "Wow. There's no way I can solve for all those variables. With four variables, I'm going to need four equations!" If so, you're absolutely right. However, the Joe Bloggs answer, (E), is not the correct answer.

## Explanations for the Joe Bloggs Examples

1. (D) is correct. Suppose a scarf has a retail price of $100. Clare's Boutique sells that scarf at an everyday 20% discount or $80. During the Mother's Day sale, the store reduces the everyday price another 10% or $8 (10% of $80). So the sale price is $72, which is 72% of the retail price.

2. (D) is correct. Combine all the equations into one and you get $(w + 2x) + (2w + 3y) + (x + 3z) = 150 + 100 + 50$. If you simplify each side, you get $3w + 3x + 3y + 3z = 300$. Divide everything by 3 and you find that $w + x + y + z = 100$. Even though you can't solve for each individual variable, you can answer. The question.

## Quiz #1

Identify the Joe Bloggs answers for each of the following questions.

1. Alex drives to and from work each day along the same route. If she drives at a speed of 80 miles per hour on the way to work and she drives at a speed of 100 miles per hour on the way from work, which of the following most closely approximates her average speed in miles per hour for the round trip?

   (A) 80.0
   (B) 88.9
   (C) 90.0
   (D) 91.1
   (E) 100.0

2. For which of the following values of $n$ is $(-0.5)^n$ the greatest?

   (A) 5
   (B) 4
   (C) 3
   (D) 2
   (E) 1

3. $\left(\sqrt{3}+\sqrt{3}+\sqrt{3}\right)^2 =$

   (A) 27
   (B) 18
   (C) 9
   (D) $3\sqrt{3}$
   (E) 3

### Quiz #1 — Answers and Explanations

1. (C) is the Joe Bloggs trap answer. Your first instinct is simply to average 80 MPH and 100 MPH to get 90 MPH. If only it were that easy! The correct answer is (B). For a more complete explanation, see the practice set in Chapter 11.
2. (A) is the Joe Bloggs trap answer. When the question asks for the greatest value, your initial reaction is probably to look for the largest number. So you should be suspicious of that answer. The correct answer is (D). For an explanation of the correct answer, see the practice set in chapter 7.

3. (C) is the Joe Bloggs trap answer. The simplest-looking thing to do is square each of the $\sqrt{3}$s to get 3 + 3 + 3 = 9. Unfortunately, you can't do that. Joe Bloggs might also pick (D) because it contains a square root and that reminds him of the numbers in the problem. The correct answer is (A). For a more complete explanation of the correct answer, check the practice set in chapter 7.

## BALLPARKING

On the GMAT, you will need to do some calculations. However, the primary purpose of the GMAT is not to test your calculation abilities. In fact, you can often shortcut a problem to save some time or eliminate answers when you're unsure of how to solve a problem.

Evan when you get stuck and can't completely solve a problem, you can often eliminate some answers that are "out of the ballpark." This won't necessarily lead you to the one and only right answer (although it might), but it will help you narrow down the possible answers before you guess. Approach the question on very simple terms and try to get a rough idea of what the answer is. Then eliminate answers that aren't very close. Try this approach to eliminate some answers on this example (from Quiz #1):

1. Alex drives to and from work each day along the same route. If she drives at a speed of 80 miles per hour on the way to work and she drives at a speed of 100 miles per hour on the way from work, which of the following most closely approximates her average speed in miles per hour for the round trip?

   (A) 80.0
   (B) 88.9
   (C) 90.0
   (D) 91.1
   (E) 100.0

You know that Alex's average speed should be pretty close to 90 MPH. That narrows the possible answers to (B), (C), and (D). You also know that (A) is wrong because the average of 80 and a higher speed must be more than 80. The same reasoning lets you eliminate (E). Factor in the Joe Bloggs technique to eliminate (C) and you're down to a 50-50 guess even if you don't know the "real" solution to the problem. As we mentioned above, the correct answer is (B).

The second aspect of ballparking is saving time by approximating. Because time is precious, you don't want to waste time when you don't need to. On the GMAT, you can use approximations to make the calculations easier and faster. Whenever you see a calculation that looks difficult, try to round off the numbers to something you can handle easily. Then see which answers are "in the ballpark." If there's only one reasonably close answer, you don't need to calculate the precise answer. If there's more than one reasonably close you answer, you'll want to go back and do the calculation to find the closest answer.

Wait a minute, isn't that a waste of time? Not really. The time you spend ballparking is really short and it will save you from choosing the wrong answer by mistake. For example, you may misplace the decimal point, or accidentally multiply a fraction instead of dividing. However, you won't pick that trap answer if you already eliminated it through ballparking. Try the ballparking approach to simplify the calculation on this next example:

2. Maggie has four quart-sized bottles of water that are each partially empty. The first bottle currently contains $\frac{1}{8}$ of a quart. The second bottle currently contains $\frac{1}{6}$ of a quart. The third bottle currently contains $\frac{1}{2}$ of a quart. The fourth bottle currently contains $\frac{3}{4}$ of a quart. How many additional quarts of water does Maggie need to fill each of these bottles?

   (A) $\frac{7}{8}$

   (B) $\frac{37}{24}$

   (C) $\frac{23}{12}$

   (D) $\frac{59}{24}$

   (E) $\frac{51}{12}$

You could solve this problem precisely by adding up the missing fractions of the bottles. However, that would be extremely time-consuming. It would involve a lot of steps and you would be likely to make mistakes. If you were lucky, you would catch the mistake and simply need to spend more time redoing the problem. If you weren't so lucky, you would choose the wrong answer. Instead, you should round off the numbers using a ballparking approach. Say that Maggie needs a full quart for each of the first two bottles (rounding up a bit). She needs one-half a quart for the third bottle. Then round down the fourth bottle (to offset the earlier rounding up) and say she doesn't need any water for it. All together, she needs approximately $2\frac{1}{2}$ quarts of water to fill the four bottles. (A) is too small—less than 1 quart. (B) is also too small—between 1 (or $\frac{24}{24}$) and 2 (or $\frac{48}{24}$) quarts. (C) is also too small—less than 2 (or $\frac{24}{12}$) quarts. (D) seems about right—between 2 (or $\frac{48}{24}$) and 3 (or $\frac{72}{24}$) quarts. (E) is too big—over 4 (or $\frac{48}{12}$) quarts. The correct answer must be (D).

## Quiz #2

Use the Ballparking approach to narrow down your choices.

1. Jeff opens a savings account with a $3,000 deposit and makes no other deposits or withdrawals. The account pays interest at the rate of 6 percent, compounded monthly. How much money, rounded to the nearest dollar, is in the account at the end of 1 year?

   (A) $2,160
   (B) $3,180
   (C) $3,185
   (D) $5,160
   (E) $6,037

2. The stock of Hi Tek, Inc., is initially offered at a price of $1 per share. If the price of the stock doubles every year, approximately how many years does it take the price to reach $60 per share?

   (A) 4
   (B) 6
   (C) 8
   (D) 10
   (E) 12

3. After a performance review, Steve's salary is increased by 5 percent. After a second performance review, Steve's new salary is increased by 20 percent. This series of raises of raises is equivalent to a single raise of

   (A) 25%
   (B) 26%
   (C) 27%
   (D) 30%
   (E) 32%

## Quiz #2—Answers and Explanations

1. (C) is correct. If the account pays simple interest (compounded annually) of 6%, the interest earned would be 6% × $3,000 = $180, for a total of $3,000 + $180 = $3,180. However, the monthly compounding means that the account earns interest slightly faster because it pays "interest on the interest." So you would choose an answer that is more than $3,180 but still in that ballpark. The only possible answer is (C). For more on compound interest, see chapter 6.
2. (B) is correct. The price starts at $1 per share. After 1 year, it doubles to $2. After a second year, it doubles again to $4. Continuing this trend, the subsequent prices are: $8, $16, $32, and $64. The stock reaches $64 per share after six years. Notice that you don't get exactly $60, but $64 is the closest, so go with that and choose (B).
3. (B) and (C) are "in the ballpark." Using the Joe Bloggs concept, you know that simply adding 5% and 20% to get 25% would be too easy. So the answer is unlikely to be (A). However, it's got to be pretty close to 25%, so you can narrow it down to (B) and (C). The correct answer is (B). For a more complete explanation, see the practice set in chapter 5.

# 3

# Data Sufficiency #1

The Data Sufficiency questions give a lot of test-takers headaches. Merely figuring out what you're asked to do can be awkward. You may know the concept a question is testing, but still miss it because you were confused by the Data Sufficiency format.

In this chapter, you'll decipher what it all means and learn a simple but extremely effective approach for those Data Sufficiency questions.

## WHAT DOES IT ALL MEAN?

The first time you get a Data Sufficiency question, you will see a screen describing the directions for that type of question. It will say something similar to:

This data sufficiency problem consists of a question and two statements, labeled (1) and (2), in which certain data are given. You have to decide whether the data given in the statements are *sufficient* for answering the question. Using the data given in the statements *plus* your knowledge of mathematics and everyday facts (such as the number of days in July or the meaning of *counterclockwise*), you must indicate whether

- statement (1) ALONE is sufficient, but statement (2) alone is not sufficient to answer the question asked;
- statement (2) ALONE is sufficient, but statement (1) alone is not sufficient to answer the question asked;
- BOTH statements (1) and (2) TOGETHER are sufficient to answer the question asked, but NEITHER statement ALONE is sufficient;
- EACH statement ALONE is sufficient to answer the question asked;
- statements (1) and (2) TOGETHER are not sufficient to answer the question asked, and additional data specific to the problem are needed.

You could easily spend a large amount of time trying to understand that confusing morass of instructions. If you do that during the test, you're throwing away precious minutes that you need to answer the questions. Instead, you should understand the Data Sufficiency format forwards and backwards before you ever set foot inside a testing center. So let's break it apart and see what those directions really mean.

Your mission on a Data Sufficiency problem is to determine which statement or combination of statements gives you enough information to answer the question. Then you choose the answer choice that matches that combination. The following chart shows you what the answers mean:

| Answer Choice | Statement (1) alone | Statement (2) alone | (1) and (2) Together |
|---|---|---|---|
| A | enough | not enough | — |
| B | not enough | enough | — |
| C | not enough | not enough | enough |
| D | enough | enough | — |
| E | not enough | not enough | not enough |

These answer choices are the same for every Data Sufficiency problem. Memorize the "definition" of each answer choice so that you never need to read the directions or the text of the answer choices.

## AD OR BCE

Don't try to look at both statements at the same time and sort out which answer fits. That way lies madness. Instead, you should look at one statement at a time, determine whether you can answer the question using the statement, and eliminate the appropriate answer choices. Here's how it all works:

**Step 1** Look at statement (1). If it provides enough information to answer the question, then you know the answer is either (A) or (D). If it doesn't provide enough information, the correct answer must be (B), (C), or (E). Write down the answers you have left.

**Step 2** Look at statement (2). You have to forget what statement (1) said while you do this. If statement (2) provides enough information, choose (B) if statement (1) *was not* sufficient in Step 1, or (D), if statement (1) *was* sufficient in Step 1. If there's not enough information, choose (A) if statement (1) *was* sufficient in Step 1, or cross off (B) and go on to Step 3, if statement (1) *was not* sufficient in Step 1.

**Step 3** If necessary, look at (1) and (2) together. If the combination provides enough information, choose answer (C). Otherwise, choose answer (E).

The next examples show this approach in action.

1. How many cookies did Max eat?

   (1) Sharon ate 4 cookies, 2 fewer than Max ate.

   (2) Max and Sharon together ate 10 cookies.

**Step 1** Look at statement (1).

This provides enough information to answer the question. Sharon ate 4 cookies, so Max ate $4 + 2 = 6$ cookies. Write down "A D" as the answers you have left.

**Step 2** Look at statement (2).

This is not enough information. You don't know how many of the 10 cookies Sharon ate and how many Max ate. You need to forget the information from statement (1) while you look at statement (2). You should eliminate (D) because statement (2) didn't work, which leaves (A) as the correct answer.

2. How many marbles does Karl have?

   (1) Karl has 6 more marbles than Jennifer has.

   (2) Jennifer has 8 marbles.

**Step 1** Look at statement (1).

This isn't enough information to answer the question. You need to add 6 to something, but you don't know what that something is. Write down "B C E" as the answers you have left.

**Step 2** Look at statement (2).

This isn't enough either. Forget statement (1) during this step. Cross off (B) because statement (2) wasn't enough, and go on to Step 3.

**Step 3** Look at statements (1) and (2) together.

Now you can find that Karl has 8 + 6 = 14 marbles. Choose (C).

## Quiz #1

1. Mike buys a bicycle and a helmet for a total cost of $315. How much does the helmet cost?

   (1) The bicycle costs twice as much as the helmet.

   (2) The bicycle costs $210.

   (A) Statement (1) ALONE is sufficient, but statement (2) alone is not sufficient to answer the question asked.
   (B) Statement (2) ALONE is sufficient, but statement (1) alone is not sufficient to answer the question asked.
   (C) BOTH statements (1) and (2) TOGETHER are sufficient to answer the question asked; but NEITHER statement ALONE is sufficient.
   (D) EACH statement ALONE is sufficient to answer the question asked.
   (E) Statements (1) and (2) TOGETHER are NOT sufficient to answer the question asked, and additional data specific to the problem are needed.

2. In an apple orchard, Carol and Joe each picked some apples. Who picked more apples?

   (1) Joe picked $\frac{3}{4}$ as many apples as Carol did.

   (2) After Carol stopped picking apples, Joe continued picking apples until he had picked 15 apples.

      (A) Statement (1) ALONE is sufficient, but statement (2) alone is not sufficient to answer the question asked.
      (B) Statement (2) ALONE is sufficient, but statement (1) alone is not sufficient to answer the question asked.
      (C) BOTH statements (1) and (2) TOGETHER are sufficient to answer the question asked; but NEITHER statement ALONE is sufficient.
      (D) EACH statement ALONE is sufficient to answer the question asked.
      (E) Statements (1) and (2) TOGETHER are NOT sufficient to answer the question asked, and additional data specific to the problem are needed.

3. While trick-or-treating at a certain house, each child in a particular group received either one or two pieces of candy. How many of the children received two pieces of candy?

   (1) Of the children in the group, 25 percent received two pieces of candy.

   (2) The children in the group received a total of 15 pieces of candy at the house.

      (A) Statement (1) ALONE is sufficient, but statement (2) alone is not sufficient to answer the question asked.
      (B) Statement (2) ALONE is sufficient, but statement (1) alone is not sufficient to answer the question asked.
      (C) BOTH statements (1) and (2) TOGETHER are sufficient to answer the question asked; but NEITHER statement ALONE is sufficient.
      (D) EACH statement ALONE is sufficient to answer the question asked.
      (E) Statements (1) and (2) TOGETHER are NOT sufficient to answer the question asked, and additional data specific to the problem are needed.

4. The rents on Doug's apartment and Magda's apartment were both increased. Which tenant paid the larger dollar increase in rent?

   (1) Doug's rent increased 2 percent.

   (2) Magda's rent increased 8 percent.

   (A) Statement (1) ALONE is sufficient, but statement (2) alone is not sufficient to answer the question asked.
   (B) Statement (2) ALONE is sufficient, but statement (1) alone is not sufficient to answer the question asked.
   (C) BOTH statements (1) and (2) TOGETHER are sufficient to answer the question asked; but NEITHER statement ALONE is sufficient.
   (D) EACH statement ALONE is sufficient to answer the question asked.
   (E) Statements (1) and (2) TOGETHER are NOT sufficient to answer the question asked, and additional data specific to the problem are needed.

5. How many guitars does Rick own?

   (1) Rick owns three times as many electric guitars as acoustic guitars.

   (2) If Rick owned 6 fewer guitars, he would own only two-thirds as many guitars as he actually owns.

   (A) Statement (1) ALONE is sufficient, but statement (2) alone is not sufficient to answer the question asked.
   (B) Statement (2) ALONE is sufficient, but statement (1) alone is not sufficient to answer the question asked.
   (C) BOTH statements (1) and (2) TOGETHER are sufficient to answer the question asked; but NEITHER statement ALONE is sufficient.
   (D) EACH statement ALONE is sufficient to answer the question asked.
   (E) Statements (1) and (2) TOGETHER are NOT sufficient to answer the question asked, and additional data specific to the problem are needed.

## Quiz #1 — Answers and Explanations

1. (D) is correct. Start with statement (1). You know that the bicycle costs twice as much as the helmet and that the two together cost $315. The only possible prices are $105 for the helmet and $210 for the bicycle. You can answer the question, so narrow it down to (A) or (D). Next look at statement (2). You can simply subtract $210 from $315 to get $105 for the cost of the helmet. This also answers the question, so choose (D).

2. (A) is correct. Start with statement (1). You don't know how many apples either person picked, but you do know that Joe picked fewer than Carol did. This answers the question, so narrow the choice to (A) or (D). Next look at statement (2). You know that Joe took longer to pick his apples, but you don't know whether he picked more or fewer than Carol did. You can't answer the question, so choose (A).

3. (C) is correct. Start with statement (1). Although you know the percentage of the group that received two pieces, you don't know how many children that is. You can't answer the question, so narrow it down to (B), (C), or (E). Next look at statement (2). You can't tell how many of the 15 pieces went to "two-piece" kids and how many went to "one-piece" kids. Maybe 1 kid received two pieces and 13 kids each received 1 piece. Maybe 7 kids each received 2 pieces and 1 kid received 1 piece. You can't answer the question, so eliminate (B). Now try both statements together. If there are $x$ children, then 75% of $x$ receive 1 piece and 25% of $x$ receive 2 pieces. Given that the total number of pieces is 15, you can set up this equation: $(0.75 \times 1 \times x) + (0.25 \times 2 \times x) = 15$. You could solve this equation and find the total number of children. Then you'd be able to answer the question by finding 25% of that number. Choose (C).

4. (E) is correct. Start with statement (1). This only mentions Doug's increase, so you can't compare that to Magda's increase because you don't know their starting rents. You can't answer the question, so narrow it down to (B), (C), or (E). Next look at statement (2). This only tells you about Magda's increase. Again, you can't answer the question because you don't know their starting rents, so eliminate (B). Next look at both statements together. Even though Magda's percent increase is larger, you don't know anything about the actual dollar increases because you don't know their starting rents. This eliminates (C). You can't answer the question, so choose (E).

5. (B) is correct. Start with statement (1). This doesn't tell you anything about the total number of guitars, just about the ratio of electric to acoustic. You can't answer the question, so narrow it down to (B), (C), or (E). Next look at statement (2). If Rick owns $y$ guitars, you can set up the equation $y - 6 = \frac{2}{3}y$. You could solve this equation to find how many guitars Rick owns. Don't waste time actually solving for $y$! You can answer the question using statement (2), so choose (B).

## DON'T FIND THE ANSWER

You usually don't need to find the actual value asked for in a Data Sufficiency question. You just need to know whether you could figure it out with the information in the statements. So don't waste your time solving the problem to come up with the numbers; just setting up the problem will usually be enough.

There is one situation in which you might want to find the number: when you're not sure whether the problem is actually solvable with the information in the statement. In that case, you should set up the problem and work through it until you are sure, even if you find yourself calculating the solution.

## PRACTICE SET

1. What was the total cost to place $q$ long distance telephone calls?

   (1) Each call lasted at least 2 minutes.

   (2) The rate for long distance calls is \$0.32 per minute and $q = 7$.

   (A) Statement (1) ALONE is sufficient, but statement (2) alone is not sufficient to answer the question asked.
   (B) Statement (2) ALONE is sufficient, but statement (1) alone is not sufficient to answer the question asked.
   (C) BOTH statements (1) and (2) TOGETHER are sufficient to answer the question asked; but NEITHER statement ALONE is sufficient.
   (D) EACH statement ALONE is sufficient to answer the question asked.
   (E) Statements (1) and (2) TOGETHER are NOT sufficient to answer the question asked, and additional data specific to the problem are needed.

2. If $s$ and $t$ are positive, what is the value of $s$?

(1) $t = 2.7$

(2) $s = 3.1t$

(A) Statement (1) ALONE is sufficient, but statement (2) alone is not sufficient to answer the question asked.
(B) Statement (2) ALONE is sufficient, but statement (1) alone is not sufficient to answer the question asked.
(C) BOTH statements (1) and (2) TOGETHER are sufficient to answer the question asked; but NEITHER statement ALONE is sufficient.
(D) EACH statement ALONE is sufficient to answer the question asked.
(E) Statements (1) and (2) TOGETHER are NOT sufficient to answer the question asked, and additional data specific to the problem are needed.

3. What is the value of $x$ ?

(1) $3x + y = 12$

(2) $y = 9$

(A) Statement (1) ALONE is sufficient, but statement (2) alone is not sufficient to answer the question asked.
(B) Statement (2) ALONE is sufficient, but statement (1) alone is not sufficient to answer the question asked.
(C) BOTH statements (1) and (2) TOGETHER are sufficient to answer the question asked; but NEITHER statement ALONE is sufficient.
(D) EACH statement ALONE is sufficient to answer the question asked.
(E) Statements (1) and (2) TOGETHER are NOT sufficient to answer the question asked, and additional data specific to the problem are needed.

4. Tom spent the day baking cookies. Of the first three dozen cookies, one-third contained chocolate chips. Of the remaining cookies, one-half contain chocolate chips. How many cookies containing chocolate chips did Tom bake that day?

   (1) Tom baked five dozen cookies.

   (2) Of all the cookies Tom baked, two-fifths contained chocolate chips.

   (A) Statement (1) ALONE is sufficient, but statement (2) alone is not sufficient to answer the question asked.
   (B) Statement (2) ALONE is sufficient, but statement (1) alone is not sufficient to answer the question asked.
   (C) BOTH statements (1) and (2) TOGETHER are sufficient to answer the question asked; but NEITHER statement ALONE is sufficient.
   (D) EACH statement ALONE is sufficient to answer the question asked.
   (E) Statements (1) and (2) TOGETHER are NOT sufficient to answer the question asked, and additional data specific to the problem are needed.

5. The usual price for a bagel was reduced during a sale. How much money could one have saved by purchasing ten bagels at the sale price rather than at the usual price?

   (1) The usual price for the bagels was $0.50 per bagel.

   (2) The sale price for the bagels was $0.40 per bagel.

   (A) Statement (1) ALONE is sufficient, but statement (2) alone is not sufficient to answer the question asked.
   (B) Statement (2) ALONE is sufficient, but statement (1) alone is not sufficient to answer the question asked.
   (C) BOTH statements (1) and (2) TOGETHER are sufficient to answer the question asked; but NEITHER statement ALONE is sufficient.
   (D) EACH statement ALONE is sufficient to answer the question asked.
   (E) Statements (1) and (2) TOGETHER are NOT sufficient to answer the question asked, and additional data specific to the problem are needed.

6. How many sandwiches were sold at a certain delicatessen yesterday?

   (1) A total of 64 sandwiches were sold at the delicatessen today, 12 more than half the number sold yesterday.

   (2) The number of sandwiches sold at the delicatessen today was 40 fewer than the number sold yesterday.

      (A) Statement (1) ALONE is sufficient, but statement (2) alone is not sufficient to answer the question asked.
      (B) Statement (2) ALONE is sufficient, but statement (1) alone is not sufficient to answer the question asked.
      (C) BOTH statements (1) and (2) TOGETHER are sufficient to answer the question asked; but NEITHER statement ALONE is sufficient.
      (D) EACH statement ALONE is sufficient to answer the question asked.
      (E) Statements (1) and (2) TOGETHER are NOT sufficient to answer the question asked, and additional data specific to the problem are needed.

7. If a total of 30 puppies are displayed in the two windows of a pet store, how many of the 30 puppies are female?

   (1) $\frac{3}{4}$ of the puppies in the left window are male.

   (2) $\frac{1}{3}$ of the puppies in the right window are female.

      (A) Statement (1) ALONE is sufficient, but statement (2) alone is not sufficient to answer the question asked.
      (B) Statement (2) ALONE is sufficient, but statement (1) alone is not sufficient to answer the question asked.
      (C) BOTH statements (1) and (2) TOGETHER are sufficient to answer the question asked; but NEITHER statement ALONE is sufficient.
      (D) EACH statement ALONE is sufficient to answer the question asked.
      (E) Statements (1) and (2) TOGETHER are NOT sufficient to answer the question asked, and additional data specific to the problem are needed.

8. By what percent did the price of a particular stock increase?

   (1) The price of the stock increased by $10.

   (2) The price of the stock doubled to $20.

   (A) Statement (1) ALONE is sufficient, but statement (2) alone is not sufficient to answer the question asked.
   (B) Statement (2) ALONE is sufficient, but statement (1) alone is not sufficient to answer the question asked.
   (C) BOTH statements (1) and (2) TOGETHER are sufficient to answer the question asked; but NEITHER statement ALONE is sufficient.
   (D) EACH statement ALONE is sufficient to answer the question asked.
   (E) Statements (1) and (2) TOGETHER are NOT sufficient to answer the question asked, and additional data specific to the problem are needed.

9. What is the value of $a + b$?

   (1) $a = 7$

   (2) $a + b - 9 = 0$

   (A) Statement (1) ALONE is sufficient, but statement (2) alone is not sufficient to answer the question asked.
   (B) Statement (2) ALONE is sufficient, but statement (1) alone is not sufficient to answer the question asked.
   (C) BOTH statements (1) and (2) TOGETHER are sufficient to answer the question asked; but NEITHER statement ALONE is sufficient.
   (D) EACH statement ALONE is sufficient to answer the question asked.
   (E) Statements (1) and (2) TOGETHER are NOT sufficient to answer the question asked, and additional data specific to the problem are needed.

10. How much did a certain taxi ride cost?

    (1) The taxi ride covered 3.75 miles.

    (2) The cost for the taxi ride was $2.00 plus $0.30 for every 0.25 miles.

    (A) Statement (1) ALONE is sufficient, but statement (2) alone is not sufficient to answer the question asked.
    (B) Statement (2) ALONE is sufficient, but statement (1) alone is not sufficient to answer the question asked.
    (C) BOTH statements (1) and (2) TOGETHER are sufficient to answer the question asked; but NEITHER statement ALONE is sufficient.
    (D) EACH statement ALONE is sufficient to answer the question asked.
    (E) Statements (1) and (2) TOGETHER are NOT sufficient to answer the question asked, and additional data specific to the problem are needed.

11. How many of the trucks in a parking lot that contains 40 motor vehicles are painted red?

    (1) Of the motor vehicles in the parking lot, 20 percent are painted red.

    (2) Of the motor vehicles in the parking lot, 15 are trucks.

    (A) Statement (1) ALONE is sufficient, but statement (2) alone is not sufficient to answer the question asked.
    (B) Statement (2) ALONE is sufficient, but statement (1) alone is not sufficient to answer the question asked.
    (C) BOTH statements (1) and (2) TOGETHER are sufficient to answer the question asked; but NEITHER statement ALONE is sufficient.
    (D) EACH statement ALONE is sufficient to answer the question asked.
    (E) Statements (1) and (2) TOGETHER are NOT sufficient to answer the question asked, and additional data specific to the problem are needed.

12. Bruce, John, Linda, and Mark stand, in that order, in a straight line. If Linda stands 7 feet away from Mark, what is the distance from Bruce to John?

(1) Bruce stands 7 feet away from Linda.

(2) John stands 11 feet away from Mark.

(A) Statement (1) ALONE is sufficient, but statement (2) alone is not sufficient to answer the question asked.
(B) Statement (2) ALONE is sufficient, but statement (1) alone is not sufficient to answer the question asked.
(C) BOTH statements (1) and (2) TOGETHER are sufficient to answer the question asked; but NEITHER statement ALONE is sufficient.
(D) EACH statement ALONE is sufficient to answer the question asked.
(E) Statements (1) and (2) TOGETHER are NOT sufficient to answer the question asked, and additional data specific to the problem are needed.

13. What is the value of $x$ ?

(1) $x + y + z = 17$

(2) $x + y = 11$

(A) Statement (1) ALONE is sufficient, but statement (2) alone is not sufficient to answer the question asked.
(B) Statement (2) ALONE is sufficient, but statement (1) alone is not sufficient to answer the question asked.
(C) BOTH statements (1) and (2) TOGETHER are sufficient to answer the question asked; but NEITHER statement ALONE is sufficient.
(D) EACH statement ALONE is sufficient to answer the question asked.
(E) Statements (1) and (2) TOGETHER are NOT sufficient to answer the question asked, and additional data specific to the problem are needed.

14. The regular price for a box of Crunch-O cereal is $3.50. How much money will be saved on the purchase of 5 boxes of Crunch-O cereal if the regular price is specially reduced?

    (1) The reduced price is more than 50% of the regular price.

    (2) The reduced price is $0.75 less per box than the regular price.

    (A) Statement (1) ALONE is sufficient, but statement (2) alone is not sufficient to answer the question asked.
    (B) Statement (2) ALONE is sufficient, but statement (1) alone is not sufficient to answer the question asked.
    (C) BOTH statements (1) and (2) TOGETHER are sufficient to answer the question asked; but NEITHER statement ALONE is sufficient.
    (D) EACH statement ALONE is sufficient to answer the question asked.
    (E) Statements (1) and (2) TOGETHER are NOT sufficient to answer the question asked, and additional data specific to the problem are needed.

15. Mike bought a laptop computer for $4,000 and later sold it. For what price did Mike sell the laptop computer?

    (1) Mike sold the computer for 60% of the price he paid for it.

    (2) Mike advertised the computer in a newspaper at a price of $3,000, which was 25% more than the price for which he actually sold it.

    (A) Statement (1) ALONE is sufficient, but statement (2) alone is not sufficient to answer the question asked.
    (B) Statement (2) ALONE is sufficient, but statement (1) alone is not sufficient to answer the question asked.
    (C) BOTH statements (1) and (2) TOGETHER are sufficient to answer the question asked; but NEITHER statement ALONE is sufficient.
    (D) EACH statement ALONE is sufficient to answer the question asked.
    (E) Statements (1) and (2) TOGETHER are NOT sufficient to answer the question asked, and additional data specific to the problem are needed.

16. How old, rounded to the nearest year, was Jim in May 1989?

    (1) Jim's friend Steve, who is exactly 2 years older than Jim, turned 25 years old in 1972.

    (2) In March 1982, Jim turned 33 years old.

    (A) Statement (1) ALONE is sufficient, but statement (2) alone is not sufficient to answer the question asked.
    (B) Statement (2) ALONE is sufficient, but statement (1) alone is not sufficient to answer the question asked.
    (C) BOTH statements (1) and (2) TOGETHER are sufficient to answer the question asked; but NEITHER statement ALONE is sufficient.
    (D) EACH statement ALONE is sufficient to answer the question asked.
    (E) Statements (1) and (2) TOGETHER are NOT sufficient to answer the question asked, and additional data specific to the problem are needed.

17. Of the 3,000 cars manufactured in Factory Q last year, how many were still in operation at the end of the year?

    (1) Of all of the cars manufactured in Factory Q, 60% were still in operation at the end of last year.

    (2) A total of 48,000 cars manufactured in Factory Q were still in operation at the end of last year.

    (A) Statement (1) ALONE is sufficient, but statement (2) alone is not sufficient to answer the question asked.
    (B) Statement (2) ALONE is sufficient, but statement (1) alone is not sufficient to answer the question asked.
    (C) BOTH statements (1) and (2) TOGETHER are sufficient to answer the question asked; but NEITHER statement ALONE is sufficient.
    (D) EACH statement ALONE is sufficient to answer the question asked.
    (E) Statements (1) and (2) TOGETHER are NOT sufficient to answer the question asked, and additional data specific to the problem are needed.

18. Norman is practicing shooting free throws, alternating right-handed shots with left-handed shots. He shoots 50 free throws in this manner, takes a break, and then shoots another 50 free-throws in the same manner. How many successful free-throws did Norman shoot?

    (1) Norman successfully shot 90% of his free-throws before the break and 60% of his free-throws after the break.

    (2) Norman shot 10 more successful left-handed free-throws than he did successful right-handed free-throws.

    (A) Statement (1) ALONE is sufficient, but statement (2) alone is not sufficient to answer the question asked.
    (B) Statement (2) ALONE is sufficient, but statement (1) alone is not sufficient to answer the question asked.
    (C) BOTH statements (1) and (2) TOGETHER are sufficient to answer the question asked; but NEITHER statement ALONE is sufficient.
    (D) EACH statement ALONE is sufficient to answer the question asked.
    (E) Statements (1) and (2) TOGETHER are NOT sufficient to answer the question asked, and additional data specific to the problem are needed.

19. Of the 40 guests at a party, 20 eat a bowl of ice cream. How many of the guests at the party eat a bowl of chocolate ice cream?

    (1) Of the guests, 10 eat a bowl of vanilla ice cream.

    (2) Of the guests, 5 eat a bowl of strawberry ice cream.

    (A) Statement (1) ALONE is sufficient, but statement (2) alone is not sufficient to answer the question asked.
    (B) Statement (2) ALONE is sufficient, but statement (1) alone is not sufficient to answer the question asked.
    (C) BOTH statements (1) and (2) TOGETHER are sufficient to answer the question asked; but NEITHER statement ALONE is sufficient.
    (D) EACH statement ALONE is sufficient to answer the question asked.
    (E) Statements (1) and (2) TOGETHER are NOT sufficient to answer the question asked, and additional data specific to the problem are needed.

20. The toll for trucks along a certain toll road is $100 plus $0.50 per pound of cargo. What is the toll for a truck carrying a cargo of bananas?

    (1) The truck weighs 4,000 pounds.

    (2) The bananas weigh 800 pounds.

    (A) Statement (1) ALONE is sufficient, but statement (2) alone is not sufficient to answer the question asked.
    (B) Statement (2) ALONE is sufficient, but statement (1) alone is not sufficient to answer the question asked.
    (C) BOTH statements (1) and (2) TOGETHER are sufficient to answer the question asked; but NEITHER statement ALONE is sufficient.
    (D) EACH statement ALONE is sufficient to answer the question asked.
    (E) Statements (1) and (2) TOGETHER are NOT sufficient to answer the question asked, and additional data specific to the problem are needed.

## PRACTICE SET—ANSWERS AND EXPLANATIONS

1. (E) is correct. Start with statement (1). This tells you nothing about the cost of the calls. You can't answer the question, so narrow the possible answers down to (B), (C), and (E). Look at statement (2). Now you know the cost per minute, but you don't know how many minutes the calls lasted. You can't answer the question, so eliminate (B). Look at (1) and (2) together. You know more, but you still don't know the exact number of minutes, only that the total was at least 14 minutes. This eliminates (C).You can't answer the question, so choose (E).

2. (C) is correct. Start with statement (1). This doesn't tell you anything about $s$. You can't answer the question, so narrow the choices down to (B), (C), and (E). Look at statement (2). Alone, this statement doesn't help. You don't know $t$, so you can't solve for $s$. You can't answer the question, so eliminate (B). Look at statements (1) and (2) combined. You can plug $t = 2.7$ into the second equation to get $s = (3.1)(2.7)$. Now you can solve for $s$ and answer the question. Choose (C).

3. (C) is correct. Start with statement (1). You can't solve for $x$ because you don't have a number to plug in for $y$. Narrow the answers to (B), (C), and (E). Look at statement (2). This only tells you about $y$, not about $x$. You can't answer the question, so eliminate (B). Look at (1) and (2) together. Now you can plug $y = 9$ into the first equation and solve for $x$. You can answer the question, so choose (C).

4. (D) is correct. Start with statement (1). You know that one-third of the first 36 cookies, or 12 cookies, contain chocolate chips. Since there are 5 dozen cookies total, that leaves 24 cookies of which one-half, or 12 cookies, have chocolate chips. That's $12 + 12 = 24$ cookies containing chocolate chips. You can answer the question, so narrow your choices to (A) and (D). Look at statement (2). From the information in the question, you know that 12 of the first 36 cookies contain chocolate chips. You also know that half of the rest have chocolate chips. That is, 1 out of every 2 cookies. So start by adding three cookies, in the ratio of 1:2 chocolate chips: no chocolate chips to the number you know; 12 and 36. Now 13 of the first 38 contain chocolate chips, but that's not two-fifths. Add two more cookies to get 14 of 40, then two more to get 15 of 42, and so on until you get to 24 of 60, which is two-fifths. You can answer the question, so choose (D).
5. (C) is correct. Start with statement (1).You don't know anything about the sale price, so you can't determine how much money is saved. You can't answer the question; narrow it down to (B), (C), and (E). Look at statement (2). Now you know the sale price, but not the regular price. You can't find the difference to answer the question, so eliminate (B). Look at statements (1) and (2) together. You know that you save $\$0.50 - \$0.40 = \$0.10$ per bagel. If you buy 10 bagels, that's $1.00 saved. You can answer the question; choose (C).
6. (A) is correct. Start with statement (1). If $x$ is the number of sandwiches sold yesterday, you can set up the equation $64 = \frac{1}{2}x + 12$. You don't actually need to solve the equation, just know that you could find $x$. You can answer the question, so narrow the choices down to (A) and (D). Look at statement (2). You could set up an equation, $y = x - 40$, but you can't solve it because you don't know $y$, the number of sandwiches sold today. You can't answer the question with only (B), so choose (A).
7. (E) is correct. Start with statement (1). You can turn the statement around to say that $\frac{1}{4}$ of the puppies in the left window are female, but you don't know anything about the proportion of females in the right window. You can't answer the question so far, so you've got answers (B), (C), and (E) left. Look at statement (2). Same problem, just with the other window. Again you can't answer the question, so eliminate (B). Look at (1) and (2) together. Although you know the male-to-female proportion in each window, you don't know how many puppies are in each window. If there are 24 puppies in the left window and 6 in the right, then there are $6 + 2 = 8$ female puppies. If there are 12 puppies in the left window and 18 puppies in the right, then there are $3 + 6 = 9$ female puppies. Don't assume the puppies are split evenly. You can't answer the question, so choose (E).

8. (B) is correct. Start with statement (1).You know the dollar amount of the increase, but you don't know the original price, so you can't find the percentage increase. You can't answer the question, so narrow it down to (B), (C), and (E). Look at statement (2). This tells you that the old price was \$10 and the new price is \$20. That's a 100% increase. You can answer the question, so choose (B).
9. (B) is correct. Start with statement (1). This only gives you information about $a$. You can't answer the question, so narrow your choices to (B), (C), and (E). Look at statement (2). You can solve for $a + b = 9$, which answers the question. You don't know what $a$ and $b$ are individually, but you don't need to, to answer the question. Choose (B).
10. (C) is correct. Start with statement (1). This tells you the length of the trip, but not how to get the cost from that. You can't answer the question, so your choices are now (B), (C), and (E). Look at statement (2). This tells you the formula for determining the cost, but you don't know how many miles long the trip was. Remember, you need to forget statement (1) at this point. Now look at statements (1) and (2) together. You can take the miles from (1) and plug them into the formula from (2) to find the cost. You don't need to actually find the answer, just to be certain that you could. Choose (C).
11. (E) is correct. Start with statement (1). You know that overall 20% are painted red, but you don't know how many trucks there are or if the trucks are different than the rest of the motor vehicles. They may be more or less likely to be painted red. You can't answer the question, so narrow the choices to (B), (C), and (E). Look at statement (2). You now know that there are 15 trucks, but you don't know anything about the percentage that are painted red. You can't answer the question; eliminate (B). Look at (1) and (2) together. There are 15 trucks and 20% of all motor vehicles are painted red, but you can't assume that the 20% statement applies to trucks specifically. Perhaps none of the trucks are red and all the red motor vehicles are non-trucks. You can't answer the question, so choose (E).
12. (C) is correct. Start by drawing a picture. It should look something like this:

Now look at statement (1). From the question, you know that LM is 7 feet. This statement tells you that BJ plus JL is also 7 feet. But you don't know how much of the 7 feet is BJ. You can't answer the question, so narrow your potential answers to (B), (C), and (E). Look at statement (2). This tells you that JM is 11 feet, which means that JL is 4 feet because, from the question, LM is 7 feet. However, you don't know anything about B or how far away he is from the others. You can't answer the question, so eliminate (B). Look at statements (1) and (2) together. From the question, you know that LM = 7. From (1) you know that BJ + JL = 7. From (2) you know that JL = 4. Substitute that into BJ + JL = 7 and you find that BJ = 3. That answers the question, so choose (C).

13. (E) is correct. Start with statement (1). You can't solve for $x$ because you don't know the values of $y$ and $z$. You can't answer the question, so narrow your choices to (B), (C), and (E). Look at statement (2). You can't solve for $x$ because you don't know what value to use for $y$. Eliminate (B). Try statements (1) and (2) together. You can solve for $z$ by substituting $x + y = 11$ into the first equation, but you can't solve for $x$. You can't answer the question, so choose (E).

14. (B) is correct. Start with statement (1). Because the statement says "more than 50%" you can't tell whether it's 51%, 99%, or somewhere in-between. You can't answer the question, so narrow it down to (B), (C), and (E). Look at statement (2). This tells you that you save \$0.75 per box, or \$3.75 for 5 boxes. That answers the question, so choose (B).

15. (D) is correct. Start with statement (1). You just need to find 60% of \$4,000 (it's \$2,400) and that's the price for which Mike sold the laptop. You can answer the question, so narrow your choices to (A) and (D). Look at statement (2). You know that 125% of the sale price is \$3,000, so you could solve for the sale price and answer the question. Choose (D).

16. (B) is correct. Start with statement (1). Because Steve turned 25 in 1972, you can determine that Jim turned 23 in 1972. So Jim must have turned 40 in 1989, but you don't whether he was 39 or 40 in September. His birthday may have been later in the year. You can't answer the question, so narrow the choices to (B), (C), and (E). Look at statement (2). If Jim turned 33 in March 1982, then he turned 40 in March 1989 and he would still be 40 in December. You can answer the question, so choose (B).

17. (E) is correct. Start with statement (1). The 60% refers to all cars *ever* manufactured in Factory Q, not just those made last year. You can't answer the question, so your potential answers narrow to (B), (C), and (E). Look at statement (2). Again, the 48,000 refers to all cars ever manufactured in Factory Q. You don't know how many of those are from last year's batch. You can't answer the question, so eliminate (B). Look at (1) and (2) together. You still don't have any information about last year's batch and you can't assume the 60% rate applies to them as well. You can't answer the question. Choose (E).

18. (A) is correct. Start with statement (1). Before the break, Norman made 90% of 50, or 45 free-throws. After the break, he made 60% of 50, or 30 free-throws. The total is $45 + 30 = 75$ successful free-throws. You can answer the question, so narrow your choices to (A) and (D). Look at statement (2). This statement doesn't help you because you don't know how many successful left-handed free-throws or successful right-handed free-throws he made. You can't find the total, so you can't answer the question. This eliminates D, so choose (A).

19. (E) is correct. Start with statement (1). You can determine that there are 10 other ice cream eaters, but you don't know how many of them eat chocolate ice cream. You can't answer the question. Narrow your choices to (B), (C), and (E). Look at statement (2). This leaves 15 other ice creams eaters, but you still don't know how many eat chocolate ice cream. You can't answer the question so eliminate (B). Look at statements (1) and (2) together. You may be tempted to think that 10 guests eat vanilla ice cream and 5 eat strawberry ice cream, leaving 5 to eat chocolate ice cream. However, you don't know whether any other flavors of ice cream (such as my favorite—mint chocolate chip) are available. You can't answer the question, so choose (E).

20. (B) is correct. Start with statement (1). This is not enough because you need to know the weight of the bananas. You can't answer the question, so narrow the choices to (B), (C), and (E). Look at statement (2). You can determine the toll, because it is \$100 plus $\$0.05 \times 800 = \$400$ for the cargo of bananas. You don't need to know the weight of the truck because that's not a factor in the toll. Choose (B).

# 4

# Terms of Endearment

Many of the questions in the GMAT Math section will use a lot of math vocabulary. You probably learned most of these terms in junior high and high school. However, if you're like most GMAT-takers, you haven't used them in years and so your memory of them is pretty hazy.

This chapter reviews the terms that are most popular on the GMAT. We'll look at the definitions and, more importantly, how those words are used in GMAT questions. Even if you feel comfortable with the definitions, you should work some or all of the examples to be certain you know how to apply those definitions.

## NUMBERS

It's often helpful to use a number line to describe different types of numbers and their properties. Here's an example of a number line:

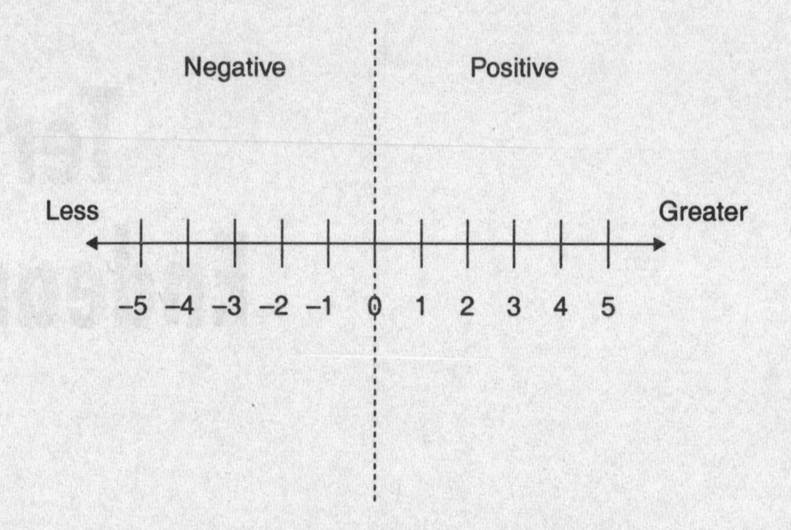


**Greater** means "to the right" on the number line, while **less** means "to the left." So 7 is greater than 5 and 2 is less than 3. Most people find that pretty intuitive. However, it's easy to get mixed up when you deal with negative numbers. For example, –5 is actually less than –3 even though 5 looks bigger than 3. You can avoid this mistake if you think in terms of "right" and "left" rather than "bigger" and "smaller."

**Positive** numbers are greater than zero. **Negative** numbers are less than zero.

**Zero** itself is neither positive nor negative. It is usually referred to as neutral.

There are some rules for multiplying (and dividing) positive and negative numbers.

| | |
|---|---|
| Positive ×/÷ Positive = Positive | $2 \times 3 = 6$ and $10 \div 5 = 2$ |
| Positive ×/÷ Negative = Negative | $2 \times (-3) = -6$ and $10 \div (-5) = -2$ |
| Negative ×/÷ Negative = Positive | $(-2) \times (-3) = 6$ and $(-10) \div (-5) = 2$ |

**Absolute value** of a number refers to that number's distance from zero. The symbol is the number between a pair of vertical lines; e.g. the absolute value of 7 is written as $|7|$. The easiest way to deal with it is simply to omit the negative sign if the number is negative. For example, $|-4| = 4$. For positive numbers, the absolute value is the same as the number itself. For example, $|3| = 3$.

**Distinct numbers** are numbers that are not equal. For example, 2 and 3 are distinct numbers, but 4 and $2^2$ are not distinct.

## Quiz #1

1. Which of the following numbers has the greatest value?

   (A) −8.3
   (B) $|-7.7|$
   (C) 2
   (D) $|4.5|$
   (E) 6.8

2. What is the value of $x$ ?

   (1) $|x| = 7$
   (2) When $x$ is divided by a negative number, the result is negative.

   (A) Statement (1) ALONE is sufficient, but statement (2) alone is not sufficient to answer the question asked.
   (B) Statement (2) ALONE is sufficient, but statement (1) alone is not sufficient to answer the question asked.
   (C) BOTH statements (1) and (2) TOGETHER are sufficient to answer the question asked; but NEITHER statement ALONE is sufficient.
   (D) EACH statement ALONE is sufficient to answer the question asked.
   (E) Statements (1) and (2) TOGETHER are NOT sufficient to answer the question asked, and additional data specific to the problem are needed.

## Quiz #1—Answers and Explanations

1. (B) is correct. The questions asks for the greatest value, so you want the number farthest to the right on a number line. The two absolute values convert to 7.7 and 4.5. Of all five numbers, 7.7 is the greatest, so (B) is the best answer.

2. (C) is correct. Start with statement (1). You know that $x = 7$ or $-7$, but that's not enough. You can't answer the question so narrow the answer down to (B), (C), or (E). Look at statement (2). From the rules for multiplying and dividing positive and negative numbers, this statement tells you that $x$ is positive, but you don't have a single value. You can't answer the question so eliminate (B). Look at (1) and (2) together. You know that $x$ is positive so the only possible value is 7. Choose (C).

## INTEGERS

An **integer** is what you commonly think of as a positive or negative "counting number." All of the numbers marked on the number line are integers. The numbers between the marks are not integers. More formally, integers include all positive whole numbers, all negative whole numbers, and 0. For example, numbers such as 1, 2, 3, 0, –1, and – 2 are all integers. Numbers such as $\frac{1}{2}$, 0.072, $-\frac{2}{3}$, and $\sqrt{3}$ are not integers.

**Even integers** are divisible by 2. **Odd integers** are not divisible by 2. It's important to realize that the terms "even" and "odd" apply only to integers. Note that zero is an even integer. There are some rules for multiplying (but not dividing) even and odd integers. If you forget these rules, you can recreate them by trying out some simple numbers.

| | |
|---|---|
| Even × Even = Even | $2 \times 4 = 8$ |
| Even × Odd = Even | $2 \times 3 = 6$ |
| Odd × Odd = Even | $3 \times 5 = 15$ |

There are also some rules for adding and subtracting odd and even integers.

| | | |
|---|---|---|
| Even +/– Even = Even | $2 + 4 = 6$ | $8 - 2 = 6$ |
| Even +/– Odd = Odd | $2 + 3 = 5$ | $8 - 3 = 5$ |
| Odd +/– Odd = Even | $3 + 5 = 8$ | $7 - 5 = 2$ |

**Consecutive** numbers are integers that follow one another. For example, 1, 2, 3,... is a series of consecutive numbers. The GMAT may also ask questions about consecutive even numbers, such as 2, 4, 6, 8,..., or consecutive odd numbers, such as 3, 5, 7, 9,...

## Quiz #2

1. If $m$ and $n$ are negative integers, which of the following must be true?

   I. $m + n < 0$

   II. $mn > 0$

   III. $mn > n$

   (A) I only
   (B) II only
   (C) I and II
   (D) I and III
   (E) I, II, and III

2. If $v$ and $w$ are both odd integers, which of the following could be an even integer?

   (A) $vw$
   (B) $v + w + 1$
   (C) $\frac{v}{w}$
   (D) $2(v + w)$
   (E) $2v + w$

3. If $a$, $b$, and $c$ are consecutive integers, which of the following must be an odd integer?

   (A) $a + b + c$
   (B) $abc$
   (C) $a + b + c + 1$
   (D) $ab(c - 1)$
   (E) $abc - 1$

## Quiz #2—Answers and Explanations

1. (E) is correct. Adding two negative numbers will just result in an even smaller (further to the left) negative number. So $m + n < 0$ and I is true. You know that a negative multiplied by a negative results in a positive number. So $mn > 0$ and II is true. This also means that $mn$ is greater than any negative number, including $n$. Therefore $mn > n$ and III is true. So the answer is (E).

2. (D) is correct. You know that an odd integer multiplied by an odd integer results in an odd integer. So $vw$ is odd and you can eliminate (A). An odd integer plus an odd integer results in an even integer. So $v + w$ is even, but $v + w + 1$ is odd and you eliminate (B). Dividing an odd integer by an odd integer gives you either an odd integer, such as $\frac{15}{5} = 3$, or a non-integer, such as $\frac{17}{5} = 3\frac{2}{5}$. So eliminate (C). An even times an odd is even, so $2v$ is even. Adding an even to an odd is odd, so $2v + w$ is odd and you should cross off (E). As we saw earlier, $v + w$ is even and multiplying by 2 keeps it even. (Even if $v + w$ were odd, multiplying by 2 would make it even.) So (D) is the best answer.

3. (E) is correct. In (A), you don't know whether you have two odd integers and one even integer (in which case $a + b + c$ is even) or two even integers and one odd integer (in which case $a + b + c$ is odd). So eliminate (A). You can also eliminate (C) because adding 1 will turn an even integer into an odd integer and vice versa, so you still don't know which one you have. In (B), $abc$ will always be even because at least one of the numbers is even and multiplying by an even number always results in an even number. So eliminate (B). In (D), either $a$ or $b$ must be even, so $ab$ must be even, which means that $ab(c - 1)$ will also be even. Eliminate (D). In (E), we know that $abc$ will always be even, so subtracting 1 will always be odd. Choose (E).

## MANIPULATING NUMBERS

**Sum** means to add. **Difference** means to subtract. **Product** means to multiply. **Quotient** means to divide.

Whenever you see one of these terms in a problem, you know to perform that operation (add, subtract, multiply, divide).

A **remainder** is the leftover from division. This is the way you did division in third grade, no decimals or fractions allowed. Just keep dividing until you get down to something smaller than the number you're dividing by. That something is the remainder.

For example, let's divide 7 by 2. Set up the long division like this: $2\overline{)7}$. You know that 2 goes into 7 three times (because $2 \times 3 = 6$).

However, there's still 1 left over. 1 is smaller than 2, so you can't continue without getting into decimals. That means 1 is the remainder.

**Reciprocal** means to flip the fraction over. For example, the reciprocal of $\frac{2}{5}$ is $\frac{5}{2}$. How do you find the reciprocal of a non-fraction such as 4? Well, first you turn 4 into a fraction by putting it over 1 to get $\frac{4}{1}$. Then it's easy to flip it over to get the reciprocal of 4, which is $\frac{1}{4}$.

## QUIZ #3

1. When 12 is divided by the positive integer $k$, the remainder is $k - 3$. Which of the following could be the value of $k$?

   (A) 3
   (B) 4
   (C) 6
   (D) 9
   (E) 10

2. If the product $xy$ is negative, which of the following must be true?

   (A) $x < 0$
   (B) $y < 0$
   (C) $\frac{x}{y} > 0$
   (D) $\frac{x}{y} < 0$
   (E) $x + y < 0$

| 5 | $t$ | 16 | 2 |
|---|---|---|---|
| 16 | | | |

3. In the figure above, the product of the two numbers in the vertical column equals the sum of the four numbers in the horizontal row. What is the value of $t$ ?

   (A) 0.5
   (B) 3
   (C) 21
   (D) 57
   (E) 80

## Quiz #3—Answers and Explanations

1. (A) is correct. For this question, just try each answer choice out and see what the remainder is. If $k = 3$ then $12 \div 3 = 4$ with a remainder of 0. Because $k - 3 = 0$, answer (A) is correct. In answer (B), the remainder is 0 and $k - 3 = 1$; so eliminate it. In answer (C), the remainder is 0 and $k - 3 = 3$; so cross it off. In answer (D), the remainder is 3 and $k - 3 = 6$; so get rid of it. In answer (E), the remainder is 2 and $k - 3 = 7$, so that's not right either.

2. (D) is correct. Because $xy$ is negative, you know that either $x$ is negative and $y$ is positive or vice versa. However, you don't know which way it is, so eliminate (A) and (B). The rule tells you that division with a positive and a negative results in a negative. Eliminate (C) and choose (D). You can cross off (E) because $x + y$ could be either positive or negative. For example, $3 + (-5) = -2$ but $(-3) + 5 = 2$. It just depends on the particular numbers you use for $x$ and $y$.
3. (D) is correct. Product means multiply, so the product of the numbers in the vertical column is $5 \times 16 = 80$. The sum of the horizontal numbers needs to equal 80 as well. Sum means add, so the sum is $5 + t + 16 + 2 = 23 + t$. If $23 + t = 80$, then $t = 57$. Choose (D).

## ZERO

Zero is a special number because it is unique in a lot of ways. Some of the trickier questions require that you be aware of some of these special characteristics. Zero is an integer, but it is neither positive nor negative. It's usually referred to as neutral. However, zero does qualify as even.

You can't divide anything by zero. Division by zero is said to be undefined. Multiplying anything by zero is always zero.

## FACTORS, MULTIPLES, AND DIVISIBILITY

**Factors** are numbers you can multiply together to get a particular number. For example, $3 \times 6 = 18$, so 3 and 6 are factors of 18. It's best to identify factors in pairs. Also, when you're trying to find all the factors of a number, start with 1 and work your way up. That way you're less likely to forget a pair.

List all the factors of 24:

$$1 \times 24$$

$$2 \times 12$$

$$3 \times 8$$

$$4 \times 6$$

When the numbers get close together, like 4 and 6, you know you've got them all.

To find the **multiples** of a number, just multiply that number by 1, then by 2, then by 3, and so on. The GMAT is generally concerned with only the positive multiples of a number. For example, the multiples of 4 are 4, 8, 12, 16, 20, 24, and so on. Don't worry about negative numbers or zero even though they technically can be multiples of a given number. There are an infinite number of multiples for any given number.

A number is **divisible** by another number if the first number divided by the second number results in an integer. For example, $8 \div 2 = 4$, so 8 is divisible by 2. However, $10 \div 4 = 2.5$, so 10 is not divisible by 4.

Here are some shortcuts to help you determine divisibility:

- A number is divisible by 2 if it is an even integer.
- A number is divisible by 3 if the sum of its digits is a number divisible by 3. For example, see whether 108 is divisible by 3. The sum of the digits is $1 + 0 + 8 = 9$. Since 9 is divisible by 3, the number 108 is also divisible by 3.
- A number is divisible by 4 if the number formed by its last two digits is divisible by 4. For example, see whether 624 is divisible by 4. The number formed by the last two digits is 24 (just ignore all of the digits except the last two). Since 24 is divisible by 4, the number 624 is also divisible by 4.
- A number is divisible by 5 if it ends in 0 or 5.
- A number is divisible by 9 if the sum of its digits is a number divisible by 9. This is very similar to the "divisible by 3" rule. For example, see whether 902,178 is divisible by 9. The sum of the digits is $9 + 0 + 2 + 1 + 7 + 8 = 27$. Since 27 is divisible by 9 (use the rule again if you're not sure), the number 902,178 is also divisible by 9.
- A number is divisible by 10 if it ends in 0.

## QUIZ #4

1. How many multiples of 3 are there between 10 and 90, inclusive?

   (A) 26
   (B) 27
   (C) 28
   (D) 29
   (E) 30

2. Which of the following is the least positive integer that is divisible by each of the integers from 2 through 5, inclusive?

   (A) 30
   (B) 60
   (C) 120
   (D) 180
   (E) 240

3. How many of the factors of 42 are divisible by 3?

   (A) 2
   (B) 3
   (C) 4
   (D) 6
   (E) 8

## Quiz #4 — Answers and Explanations

1. (B) is correct. The simplest way to solve this problem is to list every multiple of 3 from 10 to 90 and then count them. It sounds tedious but it doesn't really take as long as you think it will. The multiples of 3 are 12, 15, 18, 21, 24, 27, 30, 33, 36, 39, 42, 45, 48, 51, 54, 57, 60, 63, 66, 69, 72, 75, 78, 81, 84, 87, and 90. You include 90 because the question said *inclusive*. That's a total of 27 numbers.

2. (B) is correct. The correct answer must be divisible by 2, 3, 4, and 5. You may think the easiest way is simply to multiply those numbers together, but that's a trap. Instead, start with the smallest answer choice and see whether it's divisible by all of those numbers. If not, try the next smallest answer choice. Use the divisibility shortcuts. 30 is divisible by 2, 3, and 5, but not by 4. Eliminate (A). 60 is divisible by 2, 3, 4, and 5. Choose (B). The trap is answer (C). If you multiply $2 \times 3 \times 4 \times 5$ you'll get 120, which is divisible by all of those numbers but is not the **least** number divisible by them.

3. (C) is correct. First, list all of the factors of 42: $1 \times 42$, $2 \times 21$, $3 \times 14$, and $6 \times 7$. Of these factors, 42, 21, 3, and 6 are all divisible by 3. That's 4 numbers, so choose (C).

# PRIME NUMBERS

The official definition for **prime numbers** goes something like: "A number is prime if it has exactly 2 distinct factors, 1 and itself." This means that the number isn't divisible by anything besides itself and the number 1. The first ten prime numbers are 2, 3, 5, 7, 11, 13, 17, 19, 23, and 29.

"Wait a second!" you say. "Isn't 1 a prime number?" Actually, no. Remember, the definition says that a prime number has 2 factors. The number 1 has only 1 factor: itself. So it doesn't qualify as a prime number. This is a mistake a lot of people make, so don't get trapped.

Another thing to note is that 2 is the only even prime number. That's because all other even numbers are divisible by 2.

The **prime factors** of a number are the prime numbers you would multiply to get that number. For example, the prime factors of 6 are 2 and 3 because $2 \times 3 = 6$. The prime factors of 24 are 2, 2, 2, and 3 because $2 \times 2 \times 2 \times 3 = 24$. You could also write this as $2^3 \times 3=24$ which is sometimes called the **prime factorization** of 24. However, if the question asked for the *distinct* prime factors of 24, they are 2 and 3. You wouldn't count the extra 2's, because distinct means "not the same."

An easy way to find the prime factors of a number is a factor tree. Find two factors of the number. Then find two factors for each factor that isn't a prime number. Stop when all you have are prime numbers. This diagram is of a factor tree for 100.

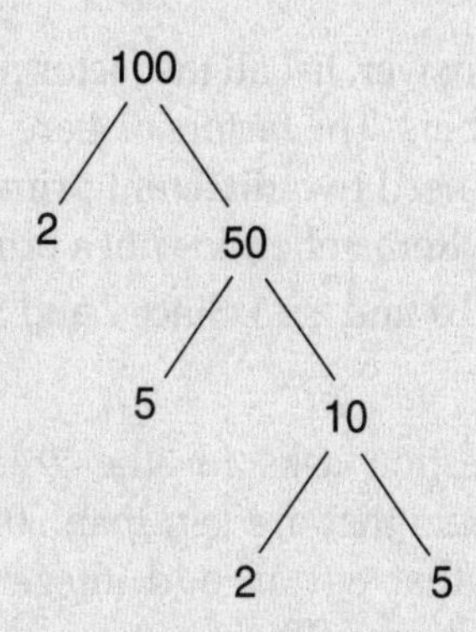


Prime factorization of 100 = 2 × 2 × 5 × 5

Distinct prime factors of 100 are 2 and 5

## Quiz #5

1. If $a$ and $b$ are distinct prime numbers, which of the following could be the product of $a$ and $b$?

   (A) 4
   (B) 5
   (C) 10
   (D) 11
   (E) 25

2. What is the greatest integer that is a sum of four different prime numbers, each less than 30?

   (A) 67
   (B) 88
   (C) 98
   (D) 104
   (E) 126

3. If $x$ is a prime number greater than 3, what is the remainder when $x^2$ is divided by 8?

   (A) 0
   (B) 1
   (C) 3
   (D) 4
   (E) 5

### Quiz #5 — Answers and Explanations

1. (C) is correct. For each answer, list all the factor pairs and see if any contain two distinct prime numbers. The factors of 4 are $1 \times 4$ and $2 \times 2$. Although 2 is a prime number, you need two **different** prime numbers. Eliminate (A). The factors of 5 are $1 \times 5$. Remember, 1 is not a prime number. Eliminate (B). The factors of 10 are $1 \times 10$ and $2 \times 5$. Since 2 and 5 are both prime numbers, choose (C).
2. (B) is correct. The question asks for the greatest integer, so add up the largest prime numbers that are less than 30. The numbers are 29, 23, 19, and 17. Remember that not all odd integers are prime. The sum is $29 + 23 + 19 + 17 = 88$. Choose (B).
3. (B) is correct. Just pick a prime number greater than 3, such as 5. Square it and you get $5^2 = 25$. Divide 25 by 8 and you get 3 with a remainder of 1. You can try out other prime numbers, but you'll always get a remainder of 1. Choose (B).

## ORDER OF OPERATIONS

The mnemonic to help you remember the order of operations is: **P E M D A S**. It stands for: **P**arentheses, **E**xponents, **M**ultiplication, **D**ivision, **A**ddition, and **S**ubtraction. This is the order you want to follow when you're solving a mathematical expression that contains more than one of these operations.

Parentheses are the first step. If you have several operations within a set of parentheses, just apply the order of operations to them. If you have parentheses inside parentheses, start with the innermost ones and work your way out.

Exponents are the next step. Handle all of the exponents before you move on to the other operations. The one exception is if you have operations within an exponent. For example, if you see $3^{2+3}$, you need to turn that into $3^5$ before you go any further.

Multiplication and division are the next step. They're really the same thing, so you won't necessarily do all the multiplication before you do division. Treat them as though they are at the same level and just work from left to right.

Addition and subtraction are the last step. Just as multiplication and division are at the same level, so are addition and subtraction. Just work from left to right.

## Quiz #6

1. $24 - 12 + \frac{36}{6} - 3 =$

   (A) 3
   (B) 5
   (C) 12
   (D) 15
   (E) 24

2. $(1 + 5) - 3^2 + 8 \div 2 \times 2$

   (A) –5
   (B) –1
   (C) 1
   (D) 5
   (E) 15

## Quiz #6 — Answers and Explanations

1. (D) is correct. There are no parentheses or exponents, so start with the division. $36 \div 6 = 6$. Then do the addition and subtraction from left to right. $24 - 12 = 12$. $12 + 6 = 18$. $18 - 3 = 15$. Choose (D).
2. (D) is correct. First, do the parentheses to get $(6) - 3^2 + 8 \div 2 \times 2$. Then do the exponent to get $6 - 9 + 8 \div 2 \times 2$. Next do the multiplication and division, left to right, to get $6 - 9 + 8$. Last do the addition and subtraction, left to right, to get 5. Choose (D).

## PRACTICE SET

1. If $x$ is a positive integer, then $x(x-1)$ is

   (A) divisible by 5 whenever $x$ is even
   (B) divisible by 9 whenever $x$ is odd
   (C) odd only when $x$ is odd
   (D) always odd
   (E) always even

2. Which of the following CANNOT result in an integer?

   (A) The product of two integers divided by the reciprocal of a different integer
   (B) An even integer divided by 7
   (C) The quotient of two distinct prime numbers
   (D) A multiple of 11 divided by 3
   (E) The sum of two odd integers divided by 2

3. If $a + b + c = 36$, what is the value of $abc$ ?

   (1) $a$, $b$, and $c$ are consecutive even integers.

   (2) $a$, $b$, and $c$ are distinct positive integers.

   (A) Statement (1) ALONE is sufficient, but statement (2) alone is not sufficient to answer the question asked.
   (B) Statement (2) ALONE is sufficient, but statement (1) alone is not sufficient to answer the question asked.
   (C) BOTH statements (1) and (2) TOGETHER are sufficient to answer the question asked; but NEITHER statement ALONE is sufficient.
   (D) EACH statement ALONE is sufficient to answer the question asked.
   (E) Statements (1) and (2) TOGETHER are NOT sufficient to answer the question asked, and additional data specific to the problem are needed.

4. How many positive integers less than 28 are prime numbers, odd multiples of 5, or the sum of a positive multiple of 2 and a positive multiple of 4?

   (A) 27
   (B) 25
   (C) 24
   (D) 22
   (E) 20

5. If a positive integer $q$ is divisible by both 3 and 11, then $q$ must also be divisible by which of the following?

   I. 14

   II. 33

   III. 66

   (A) I only
   (B) II only
   (C) III only
   (D) I and II
   (E) II and III

6. If positive integers $q$ and $r$ are both even, which of the following must be odd?

   (A) $q - r$

   (B) $\frac{q}{r}$

   (C) $\frac{q}{r} + 1$

   (D) $qr - 1$
   (E) $q(r - 1)$

7. What is the value of the two digit integer $n$ ?

   (1) $n$ is divisible by 9.

   (2) The tens digit of $n$ is 4.

   (A) Statement (1) ALONE is sufficient, but statement (2) alone is not sufficient to answer the question asked.
   (B) Statement (2) ALONE is sufficient, but statement (1) alone is not sufficient to answer the question asked.
   (C) BOTH statements (1) and (2) TOGETHER are sufficient to answer the question asked; but NEITHER statement ALONE is sufficient.
   (D) EACH statement ALONE is sufficient to answer the question asked.
   (E) Statements (1) and (2) TOGETHER are NOT sufficient to answer the question asked, and additional data specific to the problem are needed.

8. In a decreasing sequence of seven consecutive even integers, the sum of the first four integers is 68. What is the product of the last three integers in the sequence?

    (A) 1,000
    (B) 960
    (C) 925
    (D) 30
    (E) 25

9. What is the value of the integer $p$ ?

    (1) $p$ is a prime factor of 33.

    (2) $3 \leq p \leq 15$

        (A) Statement (1) ALONE is sufficient, but statement (2) alone is not sufficient to answer the question asked.
        (B) Statement (2) ALONE is sufficient, but statement (1) alone is not sufficient to answer the question asked.
        (C) BOTH statements (1) and (2) TOGETHER are sufficient to answer the question asked; but NEITHER statement ALONE is sufficient.
        (D) EACH statement ALONE is sufficient to answer the question asked.
        (E) Statements (1) and (2) TOGETHER are NOT sufficient to answer the question asked, and additional data specific to the problem are needed.

10. What is the value of the lesser of two numbers if one of the numbers is one-third the other number?

    (1) The sum of the two numbers is 16.

    (2) One number is 4.

    (A) Statement (1) ALONE is sufficient, but statement (2) alone is not sufficient to answer the question asked.
    (B) Statement (2) ALONE is sufficient, but statement (1) alone is not sufficient to answer the question asked.
    (C) BOTH statements (1) and (2) TOGETHER are sufficient to answer the question asked; but NEITHER statement ALONE is sufficient.
    (D) EACH statement ALONE is sufficient to answer the question asked.
    (E) Statements (1) and (2) TOGETHER are NOT sufficient to answer the question asked, and additional data specific to the problem are needed.

11. If $m$ is an odd integer and $n = 5m + 4$, which of the following could be a divisor of $n$ ?

    (A) 2
    (B) 3
    (C) 4
    (D) 5
    (E) 6

12. $(3 + 1)^2 + (5 - 4 \times 2) =$

    (A) 7
    (B) 12
    (C) 13
    (D) 18
    (E) 19

13. If $n \leq 20$, what is the value of $n$ ?

(1) $n$ is divisible by 2 and 5.

(2) $n$ has 2 distinct prime factors.

(A) Statement (1) ALONE is sufficient, but statement (2) alone is not sufficient to answer the question asked.
(B) Statement (2) ALONE is sufficient, but statement (1) alone is not sufficient to answer the question asked.
(C) BOTH statements (1) and (2) TOGETHER are sufficient to answer the question asked; but NEITHER statement ALONE is sufficient.
(D) EACH statement ALONE is sufficient to answer the question asked.
(E) Statements (1) and (2) TOGETHER are NOT sufficient to answer the question asked, and additional data specific to the problem are needed.

14. In $x$ is a member of the set {12, 15, 18, 24, 36, 45}, what is the value of $x$ ?

(1) $x$ is divisible by 6.

(2) $x$ is a multiple of 9.

(A) Statement (1) ALONE is sufficient, but statement (2) alone is not sufficient to answer the question asked.
(B) Statement (2) ALONE is sufficient, but statement (1) alone is not sufficient to answer the question asked.
(C) BOTH statements (1) and (2) TOGETHER are sufficient to answer the question asked; but NEITHER statement ALONE is sufficient.
(D) EACH statement ALONE is sufficient to answer the question asked.
(E) Statements (1) and (2) TOGETHER are NOT sufficient to answer the question asked, and additional data specific to the problem are needed.

15. What is the value of the sum of a sequence of $x$ consecutive even integers?

(1) $x = 5$

(2) The least integer in the sequence is 6.

(A) Statement (1) ALONE is sufficient, but statement (2) alone is not sufficient to answer the question asked.
(B) Statement (2) ALONE is sufficient, but statement (1) alone is not sufficient to answer the question asked.
(C) BOTH statements (1) and (2) TOGETHER are sufficient to answer the question asked; but NEITHER statement ALONE is sufficient.
(D) EACH statement ALONE is sufficient to answer the question asked.
(E) Statements (1) and (2) TOGETHER are NOT sufficient to answer the question asked, and additional data specific to the problem are needed.

16. If $x$ and $z$ are integers, how many even integers $y$ are there such that $x < y < z$?

(1) $z - x = 4$

(2) $x$ is odd.

(A) Statement (1) ALONE is sufficient, but statement (2) alone is not sufficient to answer the question asked.
(B) Statement (2) ALONE is sufficient, but statement (1) alone is not sufficient to answer the question asked.
(C) BOTH statements (1) and (2) TOGETHER are sufficient to answer the question asked; but NEITHER statement ALONE is sufficient.
(D) EACH statement ALONE is sufficient to answer the question asked.
(E) Statements (1) and (2) TOGETHER are NOT sufficient to answer the question asked, and additional data specific to the problem are needed.

17. What is the value of $q$ ?

(1) $q = \sqrt{9}$

(2) $|q| = 3$

(A) Statement (1) ALONE is sufficient, but statement (2) alone is not sufficient to answer the question asked.
(B) Statement (2) ALONE is sufficient, but statement (1) alone is not sufficient to answer the question asked.
(C) BOTH statements (1) and (2) TOGETHER are sufficient to answer the question asked; but NEITHER statement ALONE is sufficient.
(D) EACH statement ALONE is sufficient to answer the question asked.
(E) Statements (1) and (2) TOGETHER are NOT sufficient to answer the question asked, and additional data specific to the problem are needed.

18. What is the value of the three-digit integer $t$ if $t$ is divisible by 9?

(1) The tens digit and the hundreds digit of $t$ are both 7.

(2) The units digit of $t$ is less than both the tens digit and the hundreds digit.

(A) Statement (1) ALONE is sufficient, but statement (2) alone is not sufficient to answer the question asked.
(B) Statement (2) ALONE is sufficient, but statement (1) alone is not sufficient to answer the question asked.
(C) BOTH statements (1) and (2) TOGETHER are sufficient to answer the question asked; but NEITHER statement ALONE is sufficient.
(D) EACH statement ALONE is sufficient to answer the question asked.
(E) Statements (1) and (2) TOGETHER are NOT sufficient to answer the question asked, and additional data specific to the problem are needed.

19. If $x$ is a factor of positive integer $y$, then which of the following must be positive?

    (A) $x - y$
    (B) $y - x$
    (C) $2x - y$
    (D) $x - 2y$
    (E) $y - x + 1$

20. What is the remainder when $n$ is divided by 3?

    (1) $n$ is divisible by 5.

    (2) $n$ is divisible by 6.

    (A) Statement (1) ALONE is sufficient, but statement (2) alone is not sufficient to answer the question asked.
    (B) Statement (2) ALONE is sufficient, but statement (1) alone is not sufficient to answer the question asked.
    (C) BOTH statements (1) and (2) TOGETHER are sufficient to answer the question asked; but NEITHER statement ALONE is sufficient.
    (D) EACH statement ALONE is sufficient to answer the question asked.
    (E) Statements (1) and (2) TOGETHER are NOT sufficient to answer the question asked, and additional data specific to the problem are needed.

## Practice Set — Answers and Explanations

1. (E) is correct. You can eliminate (A) because some, but not all, even numbers for $x$ will result in a number divisible by 5. Same with (B). When $x$ is even, $x - 1$ is odd. When $x$ is odd, $x - 1$ is even. From the odd/even rules, you know that the product of an even integer and an odd integer will always be even. So $x(x - 1)$ will always be even. Choose (E).

2. (C) is correct. For this question, you need to check each answer to see if there is some combination of numbers that would result in an integer. In (A), try the product of 4 and 5 divided by the reciprocal of 2. That's $4 \times 5 \div \frac{1}{2} = 40$, which is an integer. Eliminate (A). In (B), try 14 divided by 7, which is 2; 2 is an integer, so eliminate (B). In (D), try 66 divided by 3. The result is 22, which is an integer. Eliminate (D). In (E), try 3 + 5 divided by 2, which is 4, an integer. Eliminate (E). In (C), one prime number divided by a different prime number will never result in an integer because a prime number is only divisible by 1 and itself. Choose (C).

3. (A) is correct. Start with statement (1). The three numbers must be 10, 12, and 14, so you could multiply them to find the product. You can answer the question, so the answer must be (A) or (D). Look at statement (2). This doesn't help because $a$, $b$, and $c$ could be any three integers that add up to 36. They could be 10, 12, and 14, but they could also be 1, 2, and 33 or many other combinations. You can't answer the question, so choose (A).

4. (D) is correct. Just list all the numbers and circle the ones that meet one or more of the criteria. The prime numbers are 2, 3, 5, 7, 11, 13, 17, 19, and 23. The odd multiples of 5 are 5, 15, and 25. The numbers that can be expressed as the sum of a positive multiple of 2 and a positive multiple of 4 are all the even numbers over 4. That gives you 22 numbers. Be sure you don't double count any of them. Choose (D).

5. (B) is correct. Numbers that are divisible by 3 and 11 include 33, 66, 99, 132, and so forth. You can eliminate I and II because not all these numbers are divisible by 14 or 66. All these numbers are divisible by 33, so choose (B).

6. (D) is correct. An even integer minus another even integer will always give an even result. Eliminate (A). An even integer divided by another even integer could be even, odd, or not an integer. Eliminate (B). You can also eliminate (C) because adding 1 changes odd to even and vice versa. (E) is wrong because this even times odd which will give an even result. $qr$ will always be even because even times even is even. So $qr - 1$ will always be odd. Choose (D).

7. (C) is correct. Start with statement (1). If $n$ is divisible by 9, then the sum of its digits is a multiple of 9. However, $n$ could be 18, 27, 36, 45, and so forth. You can't answer the question, so narrow the choices to (B), (C), and (E). Look at statement (2). By itself, this statement tells you that $n$ is in the 40s, but it could be 40, 41, 42, and so forth. You can't answer the question, so eliminate (B). Try statements (1) and (2) together. The only number in the 40s that is divisible by 9 is 45. You can answer the question, so choose (C).
8. (B) is correct. The first four integers must be 20, 18, 16, and 14. You can find them by dividing 68 by 4 to get 17. Then try out different groups of consecutive even integers until you find the ones that fit. The remaining three numbers are 12, 10, and 8. The product is $12 \times 10 \times 8 = 960$. Choose (B).
9. (E) is correct. Start with statement (1). 33 has two prime factors: 3 and 11, so you can't answer the question. Narrow the possible answers to (B), (C), and (E). Look at statement (2). $p$ could be any integer from 3 to 15. You can't answer the question, so eliminate (B). Look at (1) and (2) together. $p$ could still be either 3 or 11. You can't answer the question, so choose (E).
10. (A) is correct. Start with statement (1). The only two numbers that fit are 4 and 12. You can answer the question, so narrow the choices to (A) and (D). Look at statement (2). You know one of the numbers is 4, but you don't know if it's the larger or smaller number. You can't answer the question, so choose (A).
11. (B) is correct. If $m$ is odd, then $5m$ is also odd. Add 4 and $n$ remains odd. So eliminate all the even answers, (A), (C), and (E). $5m$ is a multiple of 5, so $5m + 4$ cannot be a multiple of 5 ($5m + 5$ is the next multiple of 5). Eliminate (D). The only possible answer is (B).
12. (C) is correct. Use the order of operations. Start with the parentheses. Within the second parentheses, do the multiplication before you do the subtraction. This all becomes $(4)^2 + (5 - 8) = 16 - (-3) = 13$. Choose (C).
13. (E) is correct. Look at statement (1). $n$ could be 10 or 20. You can't answer the question, so narrow the choices to (B), (C), and (E). Look at statement (2). $n$ could be several numbers, including 6 (prime factors 2 and 3), 10 (prime factors 2 and 5), 12 (prime factors 2 and 3), 14 (prime factors 2 and 7), 15 (prime factors 3 and 5), 18 (prime factors 2 and 3), and 20 (prime factors 2 and 5). You can't answer the question, so eliminate (B). Look at statements (1) and (2) together. $n$ could be 10 or 20. You can't answer the question, so choose (E).
14. (E) is correct. Look at statement (1). There are 4 numbers in the set that are divisible by 6: 12, 18, 24, and 36. You can't answer the question, so narrow the choices to (B), (C), and (E). Look at statement (2). There 3 numbers in the set that are divisible by 9: 18, 36, and 45. You can't answer the question, so eliminate (B). Look at (1) and (2) together. There are 2 numbers that are divisible by 6 and 9: 18 and 36. You can't answer the question, so choose (E).

15. (C) is correct. Look at statement (1). This tells you how many numbers to add up, but you don't know which numbers they are. You can't answer the question, so narrow the choices to (B), (C), and (E). Look at statement (2). You know that the sequence starts with 6, but you don't know how many numbers are in the sequence. You can't answer the question, so eliminate (B). Look at (1) and (2) together. The sequence is 6, 8, 10, 12, 14, and 16. You could easily find the sum, so choose (C).

16. (C) is correct. Look at statement (1). It tells you that there are 3 integers between $x$ and $z$. If $x$ is odd, then there are 2 even integers between $x$ and $z$. If $x$ is even, there is 1 even integer between $x$ and $z$. You can't answer the question, so narrow the choices to (B), (C), and (E). Look at statement (2). This doesn't help by itself because you don't know much greater $z$ is. You can't answer the question, so eliminate (B). Look at (1) and (2) together. Since $x$ is odd, there are 2 even integers between $x$ and $z$. You can answer the question, so choose (C).

17. (A) is correct. Start with statement (1). Since $\sqrt{9} = 3$ you can find the value of $q$ and answer the question. Narrow your choices to (A) and (D). Look at statement (2). This only tells you that $q = 3$ or $-3$. Don't let the information you learned in statement (1) affect your interpretation of statement (2). You can't answer the question with the information from (2), so choose (A).

18. (A) is correct. Start with statement (1). You know from the question that the sum of $t$'s digits is a multiple of 9. The sum of the tens and hundreds digits is 14. The only possible value for the units digit is 4, which makes $t = 774$. You can answer the question, so narrow the choices to (A) and (D). Look at statement (2). This leaves many possibilities for $t$, including 990, 972, 540, and so forth. You can't answer the question, so choose (A).

19. (E) is correct. You can eliminate (A) because $x$ could be less than $y$, which makes $x - y$ negative. That also lets you eliminate (C) and (D). If $x$ is a factor of $y$, then $x$ could equal $y$. For example, 10 is a factor of 10. So, (B) doesn't have to be positive; it could be 0. Eliminate (B). In (E), the greatest $x$ could be is equal to $y$, so $x - y + 1$ is at least 1. Choose (E).

20. (B) is correct. Start with statement (1). $n$ could be many different numbers such as 5, 10 , 15, and so forth. $\frac{5}{3}$ leaves a remainder of 2. $\frac{10}{3}$ leaves a remainder of 1. $\frac{15}{5}$ leaves a remainder of 0. You can't pin it down to one value, so you can't answer the question. Narrow the choices to (B), (C), and (E). Look at statement (2). Any number that's divisible by 6 is also going to be divisible by 3. So the remainder will always be 0. You can answer the question, so choose (B).

# 5

# Fractions, Decimals, and Percents, Oh My!

## FRACTIONS

Fractions are used to express division. The top number (the numerator) is divided by the bottom number (the denominator). Just think of the fraction bar as a division symbol. For example, $\frac{12}{3}$ means the same thing as $12 \div 3$, or 4.

Another way to understand fractions is that a fraction is a part of a whole. Simply put the part over the whole and you have the fraction. This approach is very useful for word problems that involve fractions. For example, suppose Oscar has a marble collection containing 2 blue marbles, 5 green marbles, and 8 red marbles. His total collection is 2 + 5 + 8 = 15 marbles. To find the fraction of his collection that is blue, put the part (2 blue marbles) over the whole (15 marbles total) and you find that $\frac{2}{15}$ of Oscar's marbles are blue.

### REDUCING FRACTIONS

A common mistake in doing fraction problems is failing to reduce the answer. If you solve the problem but don't see your answer among the choices, check to see whether you need to reduce the fraction.

You reduce a fraction by dividing the top (the numerator) and the bottom (the denominator) by the same number. Continue until there are no other numbers that will divide into both parts of the fraction.

Look at Oscar's marble collection again. Suppose you need to know what fraction of his marbles are green. Putting part over whole, you find that $\frac{5}{15}$ of the marbles are green. You look at the answers, but you don't see $\frac{5}{15}$ anywhere. Did you miss something? No. You just need to reduce the answer. Both 5 and 15 are divisible by 5. Divide each of them by 5 and $\frac{5}{15}$ becomes $\frac{1}{3}$. You can't divide any further, so you're done.

### ADDING AND SUBTRACTING FRACTIONS

A simple way to add or subtract fractions is by using "the bowtie." You multiply the bottom of the second fraction by the top of the first fraction and place the result on the top left. Then you multiply the bottom of the first fraction by the top of the second fraction and place the result on the top right. Then you multiply the bottoms of the fractions to get the denominator of the answer. Then you add (or subtract) the top numbers to get the numerator of the answer. It's easier than it sounds; look at this example:

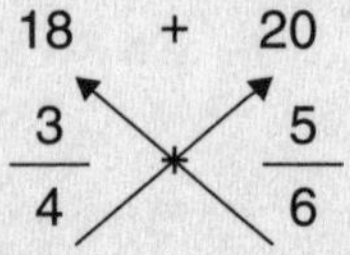

$$\begin{matrix} 18 & + & 20 \\ \frac{3}{4} & \times & \frac{5}{6} \end{matrix}$$

Multiply 6 by 3 to get 18 and put that in the upper left. Multiply 4 by 5 to get 20 and put that in the upper right. Then multiply 4 by 6 to get 24, the denominator. The criss-cross shape gives the bowtie its name. Putting it all together, you have 18 + 20 in the numerator and 24 in the denominator. That's $\frac{38}{24}$, which reduces to $\frac{19}{12}$.

Subtraction undergoes exactly the same treatment; you just subtract the top numbers instead of adding them. For example, try $\frac{2}{3}-\frac{1}{5}$. For the top numbers you get $5 \times 2 = 10$ and $3 \times 1 = 3$. For the bottom you get $3 \times 5 = 15$. Put it all together and you get $\frac{10-3}{15}=\frac{7}{15}$.

## Comparing Fractions

You can also use the bowtie to compare fractions. Suppose you need to compare $\frac{3}{5}$ to $\frac{7}{11}$ so that you can determine which is greater. Write down the fractions and use the bowtie as if you were going to add them. However, you should stop before you reach the third step (multiplying the denominators). You should have something like this:

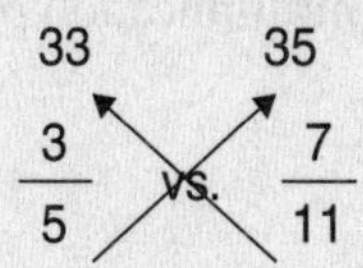


35, over the $\frac{7}{11}$, is bigger than 33, which is over the $\frac{3}{5}$, so $\frac{7}{11}$ is the greater of the two fractions.

## Multiplying Fractions

Multiplying fractions is pretty straightforward. You multiply the top numbers to get the top of the answer and you multiply the bottom numbers to get the bottom of the answer. For example, $\frac{2}{3}\times\frac{1}{5}=\frac{2\times1}{3\times5}=\frac{2}{15}$. Then you reduce if necessary.

What if you're multiplying a fraction and a whole number? Just turn the whole number into a fraction by putting a 1 in the denominator. Then multiply normally. For example, $3\times\frac{1}{5}$ becomes $\frac{3}{1}\times\frac{1}{5}=\frac{3}{5}$.

Sometimes it's helpful if you cancel the numbers before you multiply. This will make the calculation simpler. Canceling is virtually the same thing as reducing, but you do it before the calculation, not after. You can cancel any top number with any bottom number. Just divide them both by the same number. For example, $\frac{4}{5}\times\frac{3}{8}$ becomes $\frac{1}{5}\times\frac{3}{2}$ because you can divide both the 4 and the 8 by 4. Note that the answer is exactly the same regardless of whether you cancel before you multiply or reduce after you multiply. Just use whichever is more comfortable for you.

## DIVIDING FRACTIONS

To divide fractions, you multiply the first fraction by the reciprocal of the second fraction (the one you're dividing by). You just flip over the second fraction and multiply. For example, $\frac{1}{3} \div \frac{3}{4}$ becomes $\frac{1}{3} \times \frac{4}{3}$. Multiply it out and you get $\frac{4}{9}$.

Sometimes you'll see a fraction composed of two fractions, such as $\frac{\frac{3}{6}}{\frac{3}{5}}$. Just remember that you're dividing the top number by the bottom number. Rewritten, this becomes $\frac{3}{6} \div \frac{3}{5}$ or $\frac{3}{6} \times \frac{5}{3}$. Once you've written it as a multiplication problem you can cancel, but not before then. Canceling, you get $\frac{1}{8} \times \frac{5}{1} = \frac{5}{8}$.

What if the division contains a fraction and a whole number? Turn the whole number into a fraction by putting a 1 in the denominator, just as you did in multiplication. For example, $\frac{4}{15} \div 2$ becomes $\frac{4}{15} \times \frac{2}{1}$. When you flip the second fraction and multiply, you get $\frac{4}{15} \times \frac{1}{2} = \frac{2}{15}$. Don't forget either to cancel or reduce the fractions.

## MIXED FRACTIONS

A mixed fraction consists of a whole number and a fraction, such as $5\frac{1}{4}$. If a problem contains mixed fractions, the first thing you should do is convert them to "regular" fractions. Multiply the whole number (the 5 in $5\frac{1}{4}$) by the bottom of the fraction. Then add it to the top of the fraction. So $5\frac{1}{4}$ becomes $\frac{20+1}{4} = \frac{21}{4}$.

## THE REST

Word problems involving fractions can often be tricky. Look at this example:

1. Oscar has 15 marbles. If he gives $\frac{1}{3}$ of his marbles to Sally and $\frac{1}{5}$ of the rest to Mike, how many marbles does Oscar have left?

   (A) 1
   (B) 3
   (C) 5
   (D) 7
   (E) 8

Looks simple enough. One-third of his marbles is 5 marbles and one-fifth of his marbles is 3 marbles. So he gives away 5 to Sally and 3 to Mike, leaving 15 – 5 – 3 = 7 marbles for Oscar. A little too simple perhaps. Remember our friend Joe Bloggs? That's the kind of calculation he would make, and the test-writers at ETS know it. (D) is the trap answer.

Look at the problem again carefully. Oscar gives $\frac{1}{3}$ of his marbles, or 5 marbles, to Sally. He gives $\frac{1}{5}$ *of the rest* to Mike. So that's $\frac{1}{5}$ of the remaining 10, not $\frac{1}{5}$ of the original 15. Oscar gives only 2 marbles to Mike, leaving himself with 15 – 5 – 2 = 8 marbles. So the correct answer is (E). This is a very common pattern on the GMAT. Always be on the lookout for phrases such as "the rest" or "the remainder." They're there to trap the unwary.

## QUIZ #1

1. $\left(\frac{1}{2}-\frac{1}{3}\right)\times\left(\frac{2}{3}+\frac{1}{6}\right)=$

   (A) $\frac{1}{18}$

   (B) $\frac{1}{9}$

   (C) $\frac{5}{36}$

   (D) $\frac{5}{18}$

   (E) 1

2. Which of the following is greater than $\frac{3}{5}$?

   (A) $\frac{1}{2}$

   (B) $\frac{2}{7}$

   (C) $\frac{6}{11}$

   (D) $\frac{5}{9}$

   (E) $\frac{2}{3}$

3. Ian owns a collection of 60 baseball cards. If he gives $\frac{1}{5}$ of his collection to his friend Kevin and $\frac{1}{4}$ of the remainder to his friend Paul, how many baseball cards does Ian have left?

   (A) 54
   (B) 36
   (C) 33
   (D) 24
   (E) 3

## Quiz #1 — Answers and Explanations

1. (C) is correct. Use the bowtie on the first parentheses to get $\frac{1}{2}-\frac{1}{3}=\frac{3-2}{6}=\frac{1}{6}$. Bowtie the second parentheses to get $\frac{2}{3}+\frac{1}{6}=\frac{12+3}{18}=\frac{15}{18}=\frac{5}{6}$. You'll make the calculations easier if you reduce as you go. Now just multiply straight across to get $\frac{1}{6}\times\frac{5}{6}=\frac{5}{36}$. Choose (C).

2. (E) is correct. You could use the bowtie to compare all 5 answer choices to $\frac{3}{5}$. However, you'll save yourself some time if you first eliminate a few answers by Ballparking. You can eliminate (A) and (B) because $\frac{3}{5}$ is greater than $\frac{1}{2}$. You can tell because 3 is more than half of 5. Similarly, $\frac{2}{7}$ is less than $\frac{1}{2}$ so it's also less than $\frac{3}{5}$. When you bowtie $\frac{2}{3}$ and $\frac{3}{5}$, you get 10 vs. 9. The 10 goes with $\frac{2}{3}$, so choose (E).

3. (B) is correct. Ian gives $\frac{1}{5}\times 60=12$ baseball cards to Kevin, so he has $60 - 12 = 48$ left. Then he gives $\frac{1}{4}\times 48=12$ baseball cards to Paul, so he has $48 - 12 = 36$ left. Choose (B).

# DECIMALS

Decimals are a way to express numbers that are not integers. The digits to the left and right of the decimal point are referred to as decimal places. Starting at the decimal point and moving left, the places are ones (or units), tens, hundreds, thousands, and so forth. To the right of the decimal point, the places are tenths, hundredths, thousandths, and so forth. This diagram shows how it all fits together for the number 10,325.0218.

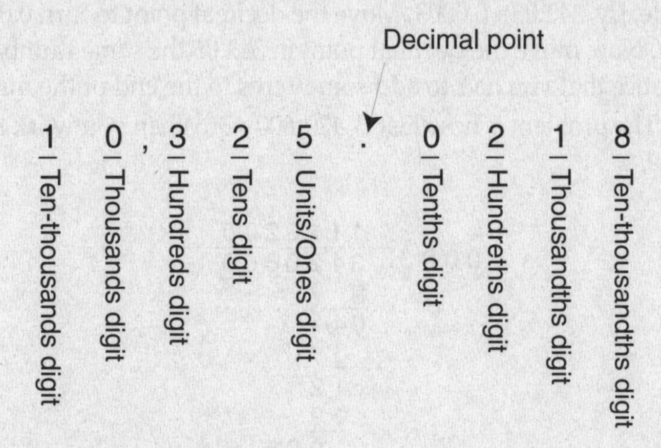


## ADDING AND SUBTRACTING DECIMALS

Adding and subtracting decimals is fairly simple. Just line up the decimals points and put a decimal point in the same spot in your answer. You may need to add some zeros to one of the numbers. For example, 0.0732 + 1.56 would look like this:

$$\begin{array}{r} 0.0732 \\ +\ 1.5600 \\ \hline 1.6332 \end{array}$$

And 7.23 – 3.105 would look like this:

$$\begin{array}{r} 7.230 \\ -\ 3.105 \\ \hline 4.125 \end{array}$$

## MULTIPLYING DECIMALS

To multiply decimals, just ignore the decimal points and multiply the numbers. Then count the number of decimal places in the original numbers. In other words, how many digits are to the right of the decimal place in the original numbers? That's the total number of decimal places you want in your answer.

For example, look at $1.05 \times 2.2$. First ignore the decimal points and just multiply. $105 \times 22 = 2310$. Then count the number of decimal places. There are two digits to the right of the decimal point in 1.05 and one digit to the right of the decimal place in 2.2. So you need three decimal places in your answer. 2310 becomes 2.310 or just 2.31. You should leave off any ending zeros.

## DIVIDING DECIMALS

To divide decimals, you have to move the decimal points in both numbers. Move the decimal point all the way to the right in the second number (the one you're dividing by) and count how many places you moved it. Then move the decimal point in the first number (the one that's being divided) the same number of places to the right. Add zeros if you run out of digits. Then simply do the long division.

As an example, try $342.06 \div 0.0003$. Move the decimal point to turn 0.0003 into 3. That's 4 decimal places. Now move the decimal point in 342.06 the same number of places (4) to get 3,420,600. Notice that you had to add some zeros to the end of the number so that you had four places. The problem is now just $3,420,600 \div 3$. When you work the long division, you get 1,140,200.

```
        1140200.
.0003 ) 342.0600
        3
        04
         3
         12
         12
          006
            6
            0
```

Here's another example: $1.175 \div 0.05$. Move the decimal point 2 places to the right in each number. That changes the problem to $117.5 \div 5$. Again, work the long division to find the answer, 23.5.

```
        23.5
.05 ) 1.17.5
      10
       17
       15
        25
        25
         0
```

## Quiz #2

1. $\frac{30.3}{4.04} =$

    (A) 0.75
    (B) 1.01
    (C) 7.5
    (D) 10.1
    (E) 75

2. $\frac{(0.15)-(0.03)}{(0.2)(0.3)} =$

    (A) 0.02
    (B) 0.12
    (C) 0.2
    (D) 2
    (E) 1.2

## Quiz #2—Answers and Explanations

1. (C) is correct. You need to move the decimal point 2 places to the right in both the top and the bottom. This gives you $\frac{3030}{404}$. Do the long division and you get 7.5. Choose (C).
2. (D) is correct. On the top, subtract the decimals by lining up the decimal point. The top is 0.12. On the bottom, multiply $2 \times 3 = 6$. There is 1 decimal place in each for a total of 2. Put 2 decimal places in the bottom and it becomes 0.06. When you divide, you need to move the decimal point 2 places to the right in each number. That gives you $12 \div 6 = 2$. Choose (D).

# PERCENTS AND CONVERSION

Percents are just like fractions and decimals; they are just another way to express non-integer numbers. The word "percent" means "for each 100." So 50% translates to "50 for each 100" or the fraction $\frac{50}{100}$, which reduces to $\frac{1}{2}$. You can also turn percents into decimals by moving the decimal point 2 places to the left. So 50% would become 0.5. After converting the percent, make the calculation using the methods you just learned for fractions and percents.

You've learned how to convert percents into fractions and decimals. You should also know how to convert the other way, fractions and decimals into percents. Last but not least, you should understand how to turn fractions into decimals and vice versa.

To convert decimals into percents, simply move the decimal point 2 places to the right. So 0.4 becomes 40% and 0.654 becomes 65.4%.

To convert fractions into percents, multiply the top and bottom of the fraction by the same number. Decide what multiplier is necessary to turn the bottom into 100. Then use that multiplier for both the top and bottom. For example, convert $\frac{3}{4}$ into a percent. You need to multiply the 4 by 25 to get 100, so that's the multiplier for both the numerator and the denominator. So you get $\frac{3}{4} = \frac{3 \times 25}{4 \times 25} = \frac{75}{100}$, which you know is the same as 75%. If you can't easily find the multiplier for the fraction (if you have a "hard" fraction such as $\frac{2}{9}$), just convert the fraction to a decimal (as described below) and then change the decimal to a percent by moving the decimal point 2 places to the right.

Converting fractions to decimals is relatively straightforward: Just divide the top number by the bottom number. So $\frac{1}{8}$ becomes 0.125.

$$\frac{1}{8} \longrightarrow \begin{array}{r} .125 \\ 8\overline{)1.000} \\ \underline{8\phantom{00}} \\ 20\phantom{0} \\ \underline{16\phantom{0}} \\ 40 \\ \underline{40} \\ 0 \end{array}$$

To convert a decimal to a fraction, first determine the place of the rightmost digit of the fraction. For example, in 0.025 the 5 is in the thousandths place. That becomes the bottom of the fraction. The top of the fraction is just the number without the decimal point. So 0.025 is the same as $\frac{25}{1000}$. Of course, you should reduce the fraction if you can, so the result is $\frac{1}{40}$.

This chart shows the conversions for the most common fractions, decimals, and percents. You should know these by heart so that you won't spend precious time calculating the converted values.

| Fraction | Decimal | Percent |
|---|---|---|
| $\frac{1}{2}$ | 0.5 | 50% |
| $\frac{1}{3}$ | $0.\overline{33}$ | $33\frac{1}{3}\%$ |
| $\frac{2}{3}$ | $0.\overline{66}$ | $66\frac{2}{3}\%$ |
| $\frac{1}{4}$ | 0.25 | 25% |
| $\frac{1}{5}$ | 0.2 | 20% |
| $\frac{1}{6}$ | $0.1\overline{66}$ | $16\frac{2}{3}\%$ |
| $\frac{1}{8}$ | 0.125 | 12.5% |

## TRANSLATION

In fraction and percent problems, especially word problems, you can get confused about which numbers to multiply together (or whatever). Translating the words into an equation is often helpful in starting a problem. Take the "stem" of the question (the part that contains the question word such as "what") and translate each word using this chart:

| Word | Translation |
|---|---|
| is, are, does (verbs) | = |
| of | × (multiply) |
| what | *n* (variable) |
| *n* percent | $\frac{n}{100}$ |

So the question "What percent of 50 is 10?" becomes $\frac{n}{100} \times 50 = 10$. Then you would solve for *n* to get the answer, which is 20%. Refer to the beginning of Chapter 7 if you need a quick refresher on solving these types of equations.

In some cases, you'll need to find numbers from the problem and insert them into this equation. Look at this next example:

1. Paul owns 7 mirrors. Of these mirrors, 3 are broken. Of the remaining mirrors, 2 have gilt frames. What fraction of Paul's unbroken mirrors have gilt frames?

(A) $\frac{1}{4}$

(B) $\frac{2}{7}$

(C) $\frac{3}{7}$

(D) $\frac{1}{2}$

(E) $\frac{4}{7}$

When you translate the question stem, you get the equation $n \times 4 = 2$. (You needed to insert $7 - 3 = 4$ for the number of unbroken mirrors.) Solving the equation you get $n = \frac{2}{4} = \frac{1}{2}$, so the answer is (D).

## Percent Change

You will probably see at least one problem in which the value of some number increases or decreases, and you need to find the "percent change." A question might also compare two numbers and ask you to state the difference in percentage terms. For both of these cases, use this formula:

$$Percent\ change = \frac{Difference}{Original\ number} \times 100$$

Finding the actual difference is usually pretty simple; just subtract the two numbers. The key is determining which number is the "original" number. Sometimes the test-writers will try to trick you. Just remember these mnemonics:

If the question mentions "increase" or "greater," you're going from a smaller number to a larger number. So the original number is the smaller one.

If the question says "decrease" or "less," you're going from a larger number to a smaller number. So the original number is the larger one.

## QUIZ #3

1. In a group of 20 tourists, 12 brought cameras. If one-half of the tourists with cameras brought disposable cameras, what percent of all the tourists brought disposable cameras?

   (A) 12%
   (B) 20%
   (C) 30%
   (D) 40%
   (E) 60%

2. Lenny can bench press 320 pounds. Ollie can bench press 400 pounds. The weight Ollie can bench press is what percent greater than the weight Lenny can bench press?

   (A) 20%
   (B) 25%
   (C) 32%
   (D) 40%
   (E) 80%

3. The original price of a model X200 laptop computer is reduced by $1,000 to the new price of $2,000. What is the percentage change in the price of the X200 laptop computer?

   (A) $12\frac{1}{2}\%$
   (B) 20%
   (C) $33\frac{1}{3}\%$
   (D) 40%
   (E) 50%

## Quiz #3—Answers and Explanations

1. (C) is correct. One-half of the 12 tourists with cameras have disposable cameras, so that's $\frac{1}{2} \times 12 = 6$ disposable cameras. There are 20 tourists, so $\frac{6}{20} = \frac{3}{10}$ of the tourists have disposable cameras. To change this fraction to a percent, multiply the top and the bottom by 10. That becomes $\frac{3 \times 10}{10 \times 10} = \frac{30}{100}$ or 30%. Choose (C).
2. (B) is correct. Divide the difference by the "original" amount. The difference is 400 – 320 = 80. To find the original amount remember that "greater" means starting small and getting big, so the original is the smaller number, 320. The percent difference is $\frac{80}{320}$ or 25%. Notice that the trap answer of 20% is there to fool you if you mistakenly divide by 400. Choose (B).
3. (C) is correct. Use the percent change formula. The difference is $1,000. The original price is $2,000 + $1,000 = $3,000. So the percent change is $\frac{1{,}000}{3{,}000} = \frac{1}{3}$ or $33\frac{1}{3}\%$. Choose (C).

# PRACTICE SET

1. In an engineering class that contains 50 students, the final exam consisted of 2 questions. Three-fifths of the students answered the first question correctly. If four-fifths of the remainder answered the second question correctly, how many students answered both questions incorrectly?

   (A) 4
   (B) 6
   (C) 10
   (D) 12
   (E) 24

2. Forty percent of $\frac{1}{8}$ of what is 20?

   (A) 400
   (B) 320
   (C) 100
   (D) 4
   (E) 1

(E) 1

3. After a performance review, Steve's salary is increased by 5 percent. After a second performance review, Steve's new salary is increased by 20 percent. This series of raises is equivalent to a single raise of

   (A) 25%
   (B) 26%
   (C) 27%
   (D) 30%
   (E) 32%

4. Of the 2,400 animals at the zoo, $\frac{1}{4}$ are primates. If the number of primates were to be reduced by $\frac{1}{4}$, what percent of the remaining animals would then be primates?

   (A) 50%
   (B) $33\frac{1}{3}\%$
   (C) 25%
   (D) 20%
   (E) 6.25%

5. If each of the following fractions were written as a decimal, which would have the fewest number of digits to the right of the decimal point?

   (A) $\frac{1}{8}$
   (B) $\frac{1}{5}$
   (C) $\frac{1}{3}$
   (D) $\frac{2}{3}$
   (E) $\frac{3}{4}$

6. $\dfrac{3\frac{1}{2}-2\frac{1}{3}}{\frac{2}{3}-\frac{3}{4}}=$

(A) $-14$

(B) $-\dfrac{1}{14}$

(C) $\dfrac{1}{14}$

(D) $1\dfrac{1}{3}$

(E) 14

7. During one day, a door-to-door brick salesman sold three-fourths of his bricks for $0.25 each. If he had 150 bricks left at the end of the day, how much money did he collect for brick sales that day?

(A) $12.50
(B) $37.50
(C) $50.00
(D) $112.50
(E) $150.00

8. In a group of 24 musicians, some are pianists and the rest are violinists. Exactly $\frac{1}{2}$ of the pianists and exactly $\frac{2}{3}$ of the violinists belong to a union. What is the least possible number of union members in the group?

(A) 12
(B) 13
(C) 14
(D) 15
(E) 16

9. Which of the following fractions is equal to the decimal 0.375?

(A) $\frac{1}{6}$

(B) $\frac{2}{7}$

(C) $\frac{3}{8}$

(D) $\frac{4}{9}$

(E) $\frac{5}{11}$

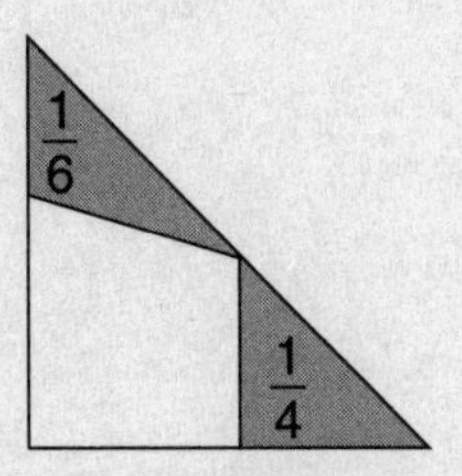


10. In the triangle ABC shown above, the two shaded regions make up $\frac{1}{4}$ and $\frac{1}{6}$ of the area of the triangle. The unshaded region makes up what fractional part of the area of the triangle?

(A) $\frac{1}{24}$

(B) $\frac{5}{12}$

(C) $\frac{7}{12}$

(D) $\frac{9}{10}$

(E) $\frac{23}{24}$

11. Justin, Max, and Paul each have a collection of marbles. Justin has 50% fewer marbles than Max has. Max has 30% more marbles than Paul has. If Paul's collection contains 80 marbles, how many marbles does Justin's collection contain?

(A) 32
(B) 48
(C) 52
(D) 56
(E) 64

12. $\dfrac{1}{2+\dfrac{2}{1+\dfrac{1}{4}}}=$

(A) $\frac{5}{18}$

(B) $\frac{5}{9}$

(C) $\frac{13}{18}$

(D) $\frac{18}{13}$

(E) $\frac{18}{5}$

13. If 12 is 20% of 40 percent of a certain number, what is the number?

(A) 20
(B) 24
(C) 72
(D) 96
(E) 150

14. The fuel efficiency of a certain make of car was increased from 30 miles per gallon for last year's model to 45 miles per gallon for this year's model. By what percent was the fuel efficiency of the car increased?

(A) 15%

(B) $33\frac{1}{3}\%$

(C) 50%

(D) $66\frac{2}{3}\%$

(E) 75%

15. $\dfrac{(4)(0.06)}{(0.12)} =$

(A) 0.18
(B) 0.2
(C) 1.8
(D) 2.0
(E) 20

16. If Kim makes a $30,000 down payment on a house that represents 20% of the sale price of the house, how much money does Kim still owe on the house?

(A) $90,000
(B) $120,000
(C) $140,000
(D) $150,000
(E) $170,000

17. If the fractions $\frac{1}{2}, \frac{2}{5}, \frac{3}{8}, \frac{4}{11}$, and $\frac{5}{7}$ were ordered from least to greatest, the second smallest number in the resulting sequence would be

(A) $\frac{1}{2}$

(B) $\frac{2}{5}$

(C) $\frac{3}{8}$

(D) $\frac{4}{11}$

(E) $\frac{5}{7}$

18. In 1991, the price of a house was 80 percent of its original price. In 1992 the price of the house was 60 percent of its original price. By what percent did the price of the house decrease from 1991 to 1992?

(A) 20%
(B) 25%
(C) $33\frac{1}{3}\%$
(D) 40%
(E) 60%

19. $1+\frac{5}{10}+\frac{3}{1{,}000}+\frac{9}{10{,}000}=$

(A) 1.539
(B) 1.5309
(C) 1.5039
(D) 1.50309
(E) 1.05039

20. During a recession, the number of ice cream cones sold in a week by a street vendor fell by $33\frac{1}{3}\%$. The vendor decided to increase the price of ice cream cones so that his weekly revenue from the cones matched the previous level. By what percent must the vendor have increased the price of ice cream cones?

    (A) 10%
    (B) 20%
    (C) 25%
    (D) $33\frac{1}{3}\%$
    (E) 50%

21. Of the 400 people in an auditorium, $\frac{1}{4}$ are wearing hats. Of those, $\frac{1}{5}$ are wearing fedoras. How many people in the auditorium are not wearing fedoras?

    (A) 20
    (B) 80
    (C) 180
    (D) 220
    (E) 380

## Practice Set—Answers and Explanations

1. (A) is correct. The number of students that answered the first question correctly is $\frac{3}{5}\times\frac{50}{1}=30$ students. That leaves 50 – 30 = 20 students who answered the first question incorrectly. Of these 20, $\frac{4}{5}\times\frac{20}{1}=16$ answered the second question correctly. That leaves 20 – 16 = 4 students who missed both questions. Choose (A).

2. (A) is correct. Use the translation technique to help you set this problem up. You get the equation $\frac{40}{100}\times\frac{1}{8}\times n=20$. Simplifying, you get $\frac{1}{20}n=20$. Multiply both sides by 20 to get $n$ = 400. Choose (A).

3. (B) is correct. The Joe Bloggs answer is (A), so you should be suspicious of that. It may help for you to make up a number for Steve's initial salary, say $100 (sorry, Steve). The 5% raise takes him to $105. The 20% raise is 20% of the new, $105 salary, not 20% of the old, $100 salary. $20\% \times \$105 = \$21$. So his new salary is $\$105 + \$21 = \$126$. That's the same as if he had gotten a raise of 26% of his original $100 salary. Choose (B). You'll see this "making up numbers" method more in Chapter 8. It's called Plugging In.

4. (D) is correct. The number of primates is $\frac{1}{4} \times 2{,}400 = 600$. The reduction in primates is $\frac{1}{4} \times 600 = 150$ primates, leaving 600 – 150 = 450 primates. Don't forget, however, that this also reduces the total number of animals at the zoo to 2,400 – 150 = 2,250. The fraction of animals at the zoo that are primates is $\frac{450}{2{,}250}$ (part over whole), which reduces to $\frac{1}{5}$. From the conversion chart, you know that's the same as 20%. Choose (D).

5. (B) is correct. Just write each of these fractions as a decimal. You probably know them from the conversion chart, but you can always use the division method if you forget. The conversions are: $\frac{1}{8} = 0.125$, $\frac{1}{5} = 0.2$, $\frac{1}{3} = 0.\overline{33}$, $\frac{2}{3} = 0.\overline{66}$, and $\frac{3}{4} = 0.75$. 0.2 is the answer because it has only one digit to the right of the decimal point. Choose (B).

6. (A) is correct. Do two subtractions first, using the bowtie, and then do the division. On the top, you'll need to convert the mixed fractions. $3\frac{1}{2}$ becomes $\frac{6+1}{2} = \frac{7}{2}$. $2\frac{1}{3}$ becomes $\frac{6+1}{3} = \frac{7}{3}$. Applying the bowtie, you get $\frac{7}{2} - \frac{7}{3} = \frac{21-14}{6} = \frac{7}{6}$ for the numerator. Applying the bowtie to the bottom, you get $\frac{2}{3} - \frac{3}{4} = \frac{8-9}{12} = -\frac{1}{12}$. To divide fractions, you flip the second one (the bottom one, in this case) and multiply. This gives you $\frac{7}{6} \times \frac{-12}{1} = -14$. Choose (A).

7. (D) is correct. The 150 bricks left is $\frac{1}{4}$ of his initial inventory of bricks, so he had $4 \times 150 = 600$ bricks. He sold 600 – 150 = 450 bricks at $0.25 each. His sales revenue was $\$0.25 \times 450 = \$112.50$. Choose (D).

8. (B) is correct. To get the least possible number of union members, you want as many pianists as possible and as few violinists as possible because $\frac{1}{2}$ is less than $\frac{2}{3}$. However, you have to have some violinists because the problem states there are some of each. It's impossible to have fractional violinists, so the number of violinists is a multiple of 3. It can't be 3 violinists and 21 pianists because that would give you $\frac{1}{2}\times 21 = 10\frac{1}{2}$ union pianists. Try 6 violinists and 18 pianists. That gives you $\frac{1}{2}\times 18 = 9$ union pianists. Combined with the $\frac{2}{3}\times 6 = 4$ union violinists, that means 9 + 4 = 13 union members total. Note that Joe Bloggs will probably pick the smallest answer when the questions asks for the "least possible number." Choose (B).

9. (C) is correct. You need to convert the fractions to decimals, either with the conversion chart or by doing the division. $\frac{1}{6} = 0.1\overline{66}$, $\frac{2}{7} = 0.2857...$, $\frac{3}{8} = 0.375$. If you want to check the other two answers (depending on where you are in the section), they are: $\frac{4}{9} = 0.\overline{44}$ and $\frac{5}{11} = 0.\overline{45}$. Choose (C).

10. (C) is correct. Use the bowtie to find the total area of the shaded regions: $\frac{1}{4}+\frac{1}{6}=\frac{6+4}{24}=\frac{10}{24}=\frac{5}{12}$. That leaves $\frac{7}{12}$ of the area for the unshaded region. Choose (C).

11. (C) is correct. Paul has 80 marbles, so Max has $30\% \times 80 = 24$ marbles more for a total of 80 + 24 = 104 marbles. Justin has $50\% \times 104 = 52$ fewer marbles than Max for a total of 104 – 52 = 52 marbles. Choose (C).

12. (A) is correct. You'll need to work this problem from the "innermost" fraction outwards. It helps if you rewrite the problem as you work each part. $1+\frac{1}{4}=\frac{4+1}{4}=\frac{5}{4}$ for the bottom. For the next step, $2\div\frac{5}{4}=2\times\frac{4}{5}=\frac{8}{5}$. Next, $2+\frac{8}{5}=\frac{10+8}{5}=\frac{18}{5}$. Last, $1\div\frac{18}{5}=1\times\frac{5}{18}=\frac{5}{18}$. Choose (A).

13. (E) is correct. Use translation to set up the equation: $12=\frac{20}{100}\times\frac{40}{100}\times n$. That simplifies to $12=\frac{2}{25}n$. Multiply both sides by $\frac{25}{2}$ to get $n = 150$. Choose (E).

14. (C) is correct. Use the percent change formula. The difference in fuel efficiency is 45 – 30 = 15. The original number was 30, so the percent change is $\frac{15}{30}=\frac{1}{2}=50\%$. The key is dividing by the right number. Choose (C).

15. (D) is correct. To get the top part of the fraction, just multiply $4 \times 6 = 24$, then add in the 2 decimal places to get 0.24. To divide by 0.12, move the decimal point two places to the right in each decimal to make it $24 \div 12 = 2$. Choose (D).

16. (B) is correct. Restate the information as "$30,000 is 20% of the sale price." Then you can translate that sentence to the equation $30{,}000 = \frac{20}{100} \times n$. Solving the equation, you find that $n$ = $150,000. However, you need to know how much she owes. That's $150,000 – $30,000 = $120,000. Choose (B).

17. (C) is correct. First, put the fraction in order. You can quickly compare fractions to $\frac{1}{2}$ by seeing the numerator is greater or less than one-half the denominator. So $\frac{5}{7}$ is more than $\frac{1}{2}$, and the other fractions are less. You need the bowtie to put the other three fractions in order. Comparing $\frac{2}{5}$ to $\frac{3}{8}$, you get $2 \times 8 = 16$ vs. $5 \times 3 = 15$, so $\frac{2}{5}$ is greater. Comparing $\frac{3}{8}$ to $\frac{4}{11}$, you get $3 \times 11 = 33$ vs. $8 \times 4 = 32$, so $\frac{3}{8}$ is greater. So the final order is $\frac{4}{11}$, $\frac{3}{8}$, $\frac{2}{5}$, $\frac{1}{2}$, and $\frac{5}{7}$. The second fraction is $\frac{3}{8}$. Choose (C).

18. (B) is correct. It's helpful to make up a number for the original sale price, say $100,000. That means the price in 1991 was $80,000 and the price in 1992 was $60,000. Use the percent change formula. The difference is $80,000 – $60,000 = $20,000. The original number is $80,000 because you're measuring only the 1991 to 1992 time period. So the percent change is $\frac{20{,}000}{80{,}000} = \frac{1}{4} = 25\%$. Choose (B).

19. (C) is correct. The easiest way to solve this problem is to convert each fraction to a decimal and then add them. $\frac{5}{10} = 0.5$, $\frac{3}{1{,}000} = 0.003$, and $\frac{9}{10{,}000} = 0.0009$. This is just like the way you convert decimals to fractions, except reversed. Adding up the numbers, you get 1 + 0.5 + 0.003 + 0.0009 = 1.5039. Choose (C).

20. (E) is correct. Once again, "Plugging In" numbers (see chapter 8) will be very helpful. Suppose the vendor normally sells \$300 in ice cream cones. The recession causes a $33\frac{1}{3}\% \times 300 = \frac{1}{3} \times 300 = \$100$ drop in revenue to \$300 – \$100 = \$200. To increase back to \$300, he needs to add \$100, which is the difference for the percent change formula. The original is \$200 because the change is from \$200 to \$300. The original is not always the first number in the problem, but the first number in that specific change. The percent change is $\frac{100}{200} = \frac{1}{2} = 50\%$. Choose (E).

21. (E) is correct. The number of people wearing hats is $\frac{1}{4} \times 400 = 100$. The number of people wearing fedoras is $\frac{1}{5} \times 100 = 20$. So the number of people not wearing fedoras is 400 – 20 = 380. Choose (E).

# 6

# Assorted Topics #1

## RATIOS

A ratio shows the relationship between two or more numbers, but it doesn't tell you the actual values of those numbers. Suppose you have a recipe for margaritas that calls for 1 ounce of tequila and 1 ounce of triple sec for every 2 ounces of lime juice. You would say that the ratio of tequila to triple sec to lime juice is 1 : 1 : 2. But the recipe alone doesn't tell you how much tequila you need for a particular batch of margaritas. For that, you also need to know some actual number, such as the total volume of the batch of margaritas.

The ratio box is an excellent tool to help you solve ratio problems. Make a grid with a row for each item in the ratio plus a row for the total. Use three columns: ratio value, multiplier, and actual value. The result will look like this:

| | Ratio Value | Multiplier | Actual Value |
|---|---|---|---|
| Item #1 | ______ | | = ______ |
| Item #2 | ______ | × ______ | = ______ |
| Item #3 | ______ | | = ______ |
| Total | ______ | | = ______ |

To use the ratio box, fill in all of the numbers from the problem. Then find the multiplier by comparing the ratio value to the actual value for one item. This multiplier will be the same for all of the items. Just multiply the ratio value by the multiplier to find all of the other actual values. Look at this example:

1. The Kosmic Kickers is a coed soccer team with 24 players on its roster. If the ratio of male players to female players is 2 to 1, how many female players are on the Kosmic Kickers' roster?

   (A) 1
   (B) 2
   (C) 8
   (D) 12
   (E) 16

First draw a ratio box and fill in all the numbers from the problem. The ratio value for male players is 2. The ratio value for female players is 1. So the ratio total is 2 + 1 = 3. The roster contains 24 players, so that number goes in the box for actual total. Here's what you've got so far:

| | Ratio | Multiplier | Actual |
|---|---|---|---|
| Male | 2 | | = |
| Female | 1 | × | = |
| Total | 3 | | = 24 |

The next step is to find the multiplier. The ratio total is 3 and the actual total is 24. So the multiplier must be 8. Write 8 in the multiplier box. If you're ever unsure of the multiplier, divide the actual total value by the ratio total value. The result is the multiplier.

To find the other actual values, multiply each ratio value by the multiplier, 8. So the actual number of male players is $2 \times 8 = 16$ and the actual number of female players is $1 \times 8 = 8$. The correct answer is (C).

## QUIZ #1

1. If the ratio of male students to female students in a philosophy class is 3 : 5, and there is a total of 40 students in the class, how many female students are in the class?

   (A) 8
   (B) 15
   (C) 16
   (D) 24
   (E) 25

2. The ratio of apples, bananas, cantaloupes, and oranges at a fruit stand is 5 : 3 : 2 : 6, respectively. The total number of these fruits is 64. If 4 cantaloupes are added to the fruit stand, what is the new ratio of bananas to cantaloupes?

   (A) 1 : 2
   (B) 2 : 3
   (C) 1 : 1
   (D) 3 : 2
   (E) 2 : 1

3. Fred has a pocketful of nickels, dimes, and quarters, in the ratio of 2 : 5 : 10, respectively. If the total value of these coins is $15.50, how many dimes are in Fred's pocket?

   (A) 17
   (B) 25
   (C) 50
   (D) 85
   (E) 155

## Quiz #1 — Answers and Explanations

1. (E) is correct. Start by drawing a ratio box like the one shown below. Then fill in the numbers from the problem (shown in bold in the diagram). In the ratio, you have 3 male students and 5 female students, for a ratio total of $3 + 5 = 8$ students. Given the actual total; 40 students, the multiplier is $40 \div 8 = 5$. Now you multiply each of the ratio values by 5 to find the actual values. The number of female students is $5 \times 5 = 25$. Choose (E).

| | Ratio | Multiplier | Actual |
|---|---|---|---|
| Male | **3** | | = 15 |
| Female | **5** | × 5 | = 25 |
| Total | 8 | | = **40** |

2. (C) is correct. Set up a ratio box with the numbers from the problem (shown in bold in the diagram below). The ratio total for the fruit is $5 + 3 + 2 + 6 = 16$. Given the actual total of 64 pieces of fruit, you can find the multiplier, which is $64 \div 16 = 4$. Now multiply all the ratio numbers by 4 to find all the actual numbers: 20 apples, 12 bananas, 8 cantaloupes, and 24 oranges. Now you add 4 cantaloupes for a new total of 12 cantaloupes. The new ratio of bananas to cantaloupes is 12 : 12, which reduces to 1 : 1. Choose (C).

| | Ratio | Multiplier | Actual |
|---|---|---|---|
| Apples | **5** | | 20 |
| Bananas | **3** | | 12 |
| Cantaloupes | **2** | × 4 | 8 |
| Oranges | **6** | | 24 |
| Totals | 16 | | **64** |

3. (B) is correct. Set up a ratio box as shown below. Notice that you need to include the value of the coins as well as the number of coins. The total ratio value of the coins is \$3.10. Given the actual total value of \$15.50, you can find the multiplier, which is $\$15.50 \div \$3.10 = 5$. Multiply all the ratio values by 5 to get the actual values. So the number of dimes is $5 \times 5 = 25$. Choose (B).

| | Ratio | Multiplier | Actual |
|---|---|---|---|
| # Nickels | **2** | | 10 |
| # Dimes | **5** | | 25 |
| # Quarters | **10** | | 50 |
| Total # | 17 | | 24 |
| | | × 5 | |
| Value of nickels | \$ 0.10 | | \$ 0.50 |
| Value of dimes | \$ 0.50 | | \$ 2.50 |
| Value of quarters | \$ 2.50 | | \$ 12.50 |
| Total value | \$ 3.10 | | **\$ 15.50** |

## PROPORTIONS

Proportions are very similar to ratios because they also show relationships between pairs of numbers. In fact, they're pretty much the same as ratios, they're just shown in a different way.

The way to solve a proportion question is to attempt to write two equal fractions. Put the units in the same place in both fractions. For example, a question might ask, "If Carol can glaze 10 ceramic pots in 2 hours, how many pots can she glaze in 3 hours?" Set up the equal fractions as shown below:

$$\frac{10 \text{ pots}}{2 \text{ hours}} = \frac{x \text{ pots}}{3 \text{ hours}}$$

Notice that the number of pots is the top of each fraction and the number of hours is the bottom of each fraction. It's important to set up your fractions in the same way. It doesn't really matter which one (pots or hours) is on top as long as it's the same in both fractions. Because you don't know the number of pots for the second fraction, use the variable $x$.

The next step is to cross-multiply the fractions. Multiply the bottom left number by the top right number and put the result on one side of the equals sign. Multiply the bottom right number with the top left number and put that on the other side of the equals sign. You end up with $10 \times 3 = 2x$. Solving the equation, you get $x = 15$, so Carol can glaze 15 pots in 3 hours. If you'd like a review of how to solve equations, check out the beginning of Chapter 7.

## QUIZ #2

1. Chris plays solitaire at the constant rate of 3 hands in 20 minutes. How many hands of solitaire can Chris play in 2 hours?

   (A) 6
   (B) 9
   (C) 12
   (D) 18
   (E) 20

2. A bottle-capping machine caps 12 bottles per minute. How many minutes does the machine take to cap 30 bottles?

   (A) 2.5
   (B) 5
   (C) 6
   (D) 12
   (E) 60

## Quiz #2—Answers and Explanations

1. (D) is correct. Set up the two equivalent fractions as shown below. Notice that you need to change 2 hours to 120 minutes. You must use the same units in both fractions. Then cross-multiply to get $3 \times 120 = 20x$, which simplifies to $360 = 20x$. Divide both sides by 20 to get $x = 18$. Chris can play 18 hands of solitaire in 2 hours. Choose (D).

$$\frac{3 \text{ hands}}{20 \text{ minutes}} = \frac{x \text{ hands}}{120 \text{ minutes}}$$

2. (A) is correct. Set up the equivalent fractions as shown below. Notice that the variable is in the bottom of the fraction. That's okay as long as the fractions are consistent. Cross-multiply to get $12x = 30 \times 1$ or simply $12x = 30$. Divide both sides by 12 to get $x = 2.5$. Choose (A).

$$\frac{12 \text{ caps}}{1 \text{ minute}} = \frac{30 \text{ caps}}{x \text{ minutes}}$$

# AVERAGES

An average tells you something about a group of people or things. The terms *mean* and *arithmetic mean* are also used to refer to averages. There are three numbers involved in any average question: the number of things in the group, the total amount, and the average amount. Given any two of these numbers, you can always find the third. The relationship of these three numbers is shown in this formula:

$$Average = \frac{Total}{\# \text{ of things}}$$

With this many vague numbers flying around, things get confusing. A tool called the *average circle* will help you keep everything organized and, as soon as you try a couple of problems, you'll see that finding averages is pretty simple. Look at the diagram below. This gives you a place to put the three numbers for an average. Draw a separate circle for each average in the problem.

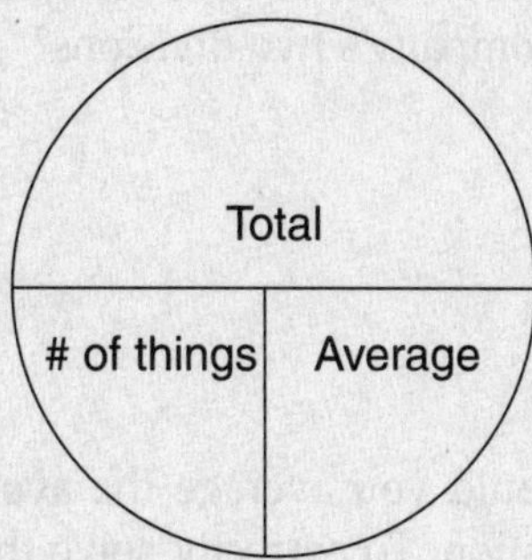


Fill in the two numbers the problem gives you and use that information to calculate the missing number. Notice that the line between the total and the other two numbers acts as the division line in a fraction. The total amount divided by the number of things gives you the average. The total amount divided by the average gives you the number of things. If you have the two bottom numbers, the number of things and the average, just multiply them to get the total amount. Try this problem:

1. Five friends play blackjack in Las Vegas and lose an average of $100 each. If the losses of two of the friends total $380, what is the average loss of the other friends?

   (A) $40
   (B) $60
   (C) $100
   (D) $120
   (E) $1,900

Draw two circles, one for the average of all five friends and one for the average of the other friends. In the first circle, fill in 5 for the number of things and $100 for the average. From this you can calculate the total loss, which is $5 \times \$100 = \$500$. For the second circle, fill in $5 - 2 = 3$ for the number of things. The question asks for the average, so you first need to find the total loss of the other friends. You know the total lost by all five friends was $500 and that the two friends lost a total of $380, so the total loss of the other three friends must be $\$500 - \$380 = \$120$. Fill that in for the total in the second circle. Now you can calculate the average loss of the three friends, which is $\frac{\$120}{3} = \$40$. The answer is (A).

Some questions will require you to combine numbers from several circles. It's okay to add and subtract the numbers of things and the total amounts, but never, ever, combine the averages directly. Averages must either be taken from the questions or calculated from the other two numbers in the circle. You can see this trap in the next example:

2. A company composed of two divisions is reviewing managerial salaries. The two managers in Division A of the company earn an average annual salary of \$75,000. The three managers in Division B of the company earn an average annual salary of \$100,000. What is the average salary of all the managers in the company's two divisions?

   (A) \$75,000
   (B) \$85,000
   (C) \$87,500
   (D) \$90,000
   (E) \$100,000

Under no circumstances should you average the averages. That answer, \$87,500, is there, but it's a trap in this question. To correctly solve the problem, set up three circles: Division A, Division B, and the whole company. In Division A, the number of things is 2, the average is \$75,000, and the total is $2 \times \$75,000 = \$150,000$. In Division B, the number of things is 3, the average is \$100,000, and the total is $3 \times \$100,000 = \$300,000$. In the whole company, the number of things is $2 + 3 = 5$, the total is $\$150,000 + \$300,000 = \$450,000$. So the average is $\frac{\$450,000}{5} = \$90,000$. The correct answer is (D).

## QUIZ #3

1. The average (arithmetic mean) of 4 numbers is 5.5. When an additional number is included, the average of all 5 numbers is 6. What is the additional number?

   (A) 5.75
   (B) 6.5
   (C) 8.0
   (D) 22.0
   (E) 30.0

2. A set of 10 numbers has an average (arithmetic mean) of 12. When a subset of numbers, which has an average of 15, is removed from the original set, the sum of the remaining numbers is 60. How many numbers remain from the original set?

   (A) 3
   (B) 4
   (C) 5
   (D) 6
   (E) 8

## Quiz #3 — Answers and Explanations

1. (C) is correct. Draw two average circles, one for the set of 4 numbers and one for the set of 5 numbers. Fill in the numbers provided in the problem (shown in bold in the diagram below). In the set of 4 numbers, the total is $4 \times 5.5 = 22$. In the set of 5 numbers, the total is $5 \times 6 = 30$. So the additional number must be $30 - 22 = 8$. Choose (C).

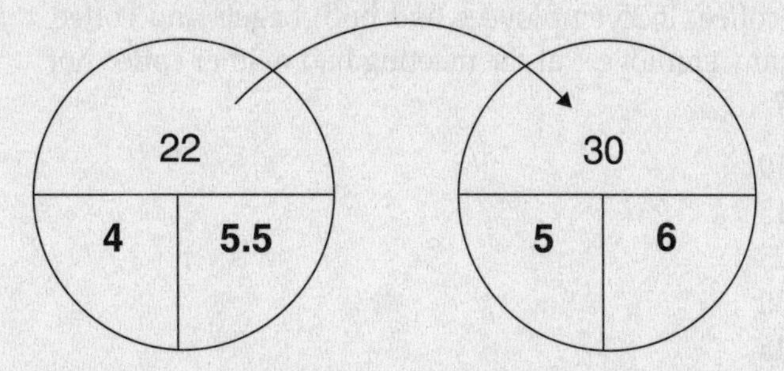


2. (D) is correct. Draw three average circles: The original set, the removed subset, and the remaining members of the original set. Fill in the numbers from the problem (shown in bold in the diagram below). In the original set, the total is $10 \times 12 = 120$. The total of the removed numbers is the difference between the totals of the original set and the remaining numbers, or $120 - 60 = 60$. Given the average of 15, the number of numbers that were removed is $60 \div 15 = 4$. So the number of remaining numbers is $10 - 4 = 6$. Choose (D).

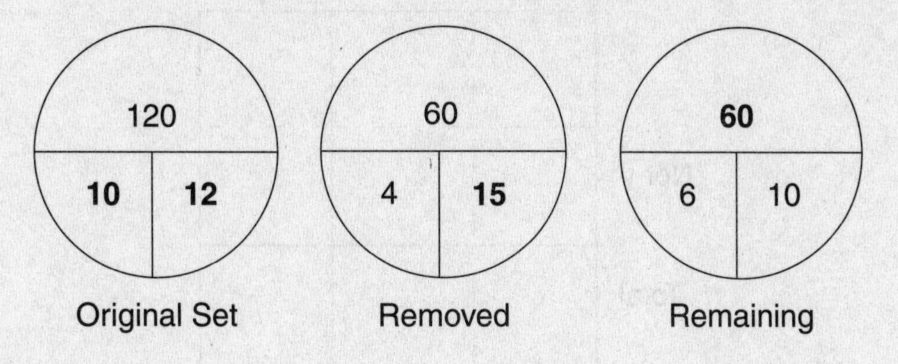


# GROUPS

Some GMAT problems deal with classifying a group of people or things into sub-groups. Many of these problems break the people into two classes with some people in both classes and some people in neither class. Although these problems look complex, you can easily solve them by applying this formula:

$$Total = Class\ 1 + Class\ 2 - Both + Neither$$

Simply plug in the numbers the problem gives you and solve for the missing value. It's possible that the problem will stipulate that everyone belongs to one class or the other. In that case, just plug in 0 for the Neither category. If the problem states that no one belongs to both classes, just use 0 for the Both value. Here's an example:

1. A company serves bagels and coffee at a breakfast meeting. Of the 120 employees at the meeting, 50 ate bagels and 95 drank coffee. If 35 employees had both bagels and coffee, how many employees at the meeting had neither coffee nor bagels?

   (A) 10
   (B) 15
   (C) 25
   (D) 30
   (E) 45

Plug the numbers into the formula and you get 120 = 50 + 95 – 35 + None. When you solve the equation, you get None = 10. The answer is (A). Look, it's much easier than it seemed at first.

Sometimes group problems get tough and the formula doesn't help. These questions will still feature two categories, but they will talk about the number of people that are and are not in each group. To solve these tough problems, you'll need to set up a grid like this one:

| | $x$ | Not $x$ | Total |
|---|---|---|---|
| $y$ | | | |
| Not $y$ | | | |
| Total | | | |

In each row, the numbers add across to give you the total on the right. In the top row, for example, the number of things that are $y$ and $x$ plus the number of things that are $y$ and not $x$ equals the total number of things that are $y$. In each column, the numbers add down to give you the totals on the bottom. In the leftmost column, for example, the number of things that are $x$ and $y$ plus the number of things that are $x$ and not $y$ equals the total number of things that are $x$. Given any two of the three numbers from a column (or a row), you can find the third number through addition or subtraction. Look at this next example:

2. Of the 200 employees at Company A, 70 work part-time and the rest work full-time. If 140 of the employees like their jobs and 10 of the part-time employees don't like their jobs, how many full-time employees like their jobs?

   (A) 50
   (B) 60
   (C) 80
   (D) 130
   (E) 140

Set up the grid as shown below. Plug in the numbers provided in the question. These are shown in bold type in the diagram. Now you can figure out the other numbers. There are 70 part-time employees out of 200 total, so there must be 130 full-time employees. Since 10 of the 70 part-time employees don't like their jobs, the other 60 must like their jobs. If 140 of the 200 employees like their jobs, the other 60 don't like their jobs. Of the 140 employees who like their jobs, 60 are part-time, so the other 80 must be full-time. Of the 130 full-time employees, 80 like their jobs, so the other 50 don't like their jobs. Every time you have two of the three numbers in a row or column, you can find the third number. The number of full-time employees who like their jobs is 80, so (C) is the correct answer.

| | Part-time | Full-time | Total |
|---|---|---|---|
| Like job | 60 | 80 | **140** |
| Don't like job | **10** | 50 | 60 |
| Total | **70** | 130 | **200** |

## QUIZ #4

1. The 2,000 students at College Q get one week for spring break. Each student takes the opportunity to travel to the beach, the mountains, or both. If 1,500 students take a trip to the beach and 700 take a trip to the mountains, how many students go to both the beach and the mountains?

   (A) 200
   (B) 500
   (C) 800
   (D) 1,300
   (E) 2,200

2. The 25 players on the Little Sluggers baseball team are supposed to bring their caps and gloves to each game. At the first game of the season, 22 players brought their caps and 18 brought their gloves. If 6 players brought their caps but not their gloves, how many players brought neither cap nor glove to the first game?

(A) 1
(B) 2
(C) 3
(D) 4
(E) 7

## Quiz #4 — Answers and Explanations

1. (A) is correct. Set up the group formula with the numbers from the problem: 2,000 = 1,500 + 700 – Both + 0. Use the 0 for None because the problem states that every student takes a trip. Simplifying the equation, you get 2,000 = 2,200 – Both, and the solution is Both = 200. Choose (A).

2. (A) is correct. Set up the grid as shown. The numbers given in the problem are shown in bold. Of the 25 players, 22 brought their caps, so 3 did not. You know 18 players brought gloves, so 7 did not. Of the 22 players with caps, 6 forgot gloves, so 16 remembered to bring gloves. Of the 18 with gloves, 16 brought caps, so 2 didn't. Of the 3 who did not bring caps, 2 brought gloves, so 1 player brought neither cap nor glove. Just where did he think he was going, anyway? Choose (A).

| | Gloves | No gloves | Total |
|---|---|---|---|
| Caps | 16 | **6** | **22** |
| No caps | 2 | 1 | 3 |
| Total | **18** | 7 | **25** |

# INTEREST

One of the few business-related topics that the GMAT tests is interest rates. To calculate the interest earned in a year, multiply the initial amount of money (also called the principal) by the percentage interest rate. For example, if $500 is deposited for one year in a savings account that earns 4% interest, then the account will earn \$500 $\times$ 4% = \$20 in interest for that year. So the total amount in the savings account after one year is \$500 + \$20 = \$520. This type of interest calculation is sometimes called simple interest.

On the GMAT, you will often need to calculate compound interest, or "interest on interest." This is necessary when money earns interest for periods longer than one year and also when the interest compounds more often than annually. For example, suppose that \$500 is deposited for ten years in a savings account earning 4% interest. In the first year, the account earns \$20 in interest, so during the second year, the account earns interest on \$520, not \$500. In the third year, the principal is even greater. Thus, the account earns interest on the interest. While this is great for your IRA account, it can make GMAT questions a little tougher.

Fortunately, you can usually avoid calculating the precise amount of compound interest. Instead, use Ballparking to find the answer. Start by finding the simple interest, which is pretty, well, simple. Then find an answer that's a little bit bigger. Look at this next example:

1. Allan deposits \$1,000 into a bank account that pays 3% interest annually. If he makes no other deposits or withdrawals, approximately how much money is in the account after 4 years?

   (A) \$120
   (B) \$126
   (C) \$1,120
   (D) \$1,126
   (E) \$1,200

Start by calculating the simple interest. The interest for one year is \$1,000 $\times$ 3% = \$30. For four years, that's \$120 in simple interest, which makes the total amount in the account \$1,000 + \$120 = \$1,120. Look for the answer that's a little bit bigger. The answer has to be (D).

Every once in a while, you may need to know the formula for compound interest. You probably won't need to make the calculation, but you may need to set up the equation. The formula is: *Total amount* = *Principal* $\times (1 + r)^t$, in which $r$ is the interest rate (expressed as a decimal) for the compounding period and $t$ is the number of compounding periods. You'll need the formula for this next example:

2. Amy deposits \$100 into an account that pays 8% interest, compounded semi-annually. She makes no other deposits or withdrawals. Which of the following expresses the amount of money the account will contain in 2 years?

   (A) $\$100 \times (1 + .04)^2$
   (B) $\$100 \times (1 + .04)^4$
   (C) $\$100 \times (1 + .04)^6$
   (D) $\$100 \times (1 + .08)^2$
   (E) $\$100 \times (1 + .08)^6$

You use 0.04 for $r$ because the account earns 4% over six months; the 8% rate is for a full year. You use 4 for $t$ because 2 years contains 4 six-month periods. So the correct answer is (B). Notice that you're not going to complete the calculation. You'd need a calculator for that and it's not necessary; just setting up the equation is enough.

## Quiz #5

1. Mark deposits $1,000 into a bank account that pays annual interest of 7%. If he makes no other deposits or withdrawals, approximately how much interest has the account earned after 4 years?

   (A) $280
   (B) $311
   (C) $700
   (D) $1,280
   (E) $1,311

2. Gilbert deposits $200 into a bank account that earns interest at the rate of 8%. He makes no further transactions other than a $100 withdrawal after 3 years. Approximately how much money is in Gilbert's account after he makes the withdrawal?

   (A) $124
   (B) $127
   (C) $148
   (D) $154
   (E) $224

## Quiz #5—Answers and Explanations

1. (B) is correct. Simple interest would be $4 \times 7\% \times \$1{,}000 = \$280$. The interest compounds (more than one year), so choose the next higher answer. Choose (B).

2. (D) is correct. Gilbert made his $100 withdrawal after it earned interest, not before then. Calculate the interest on $200, then subtract $100 from the total. Simple interest for 3 years would be $3 \times 8\% \times 200 = \$48$, for a total of $200 + 48 - 100 = \$148$. So choose the answer that's a little higher because the interest compounds. Choose (D).

## PROBLEM SET

1. Three hungry children, Sharon, Carol, and Elinor, agree to divide a batch of cookies in the ratio 3 : 5 : 7, respectively. If Sharon's share was 15 cookies, how many cookies were in that batch?

   (A) 60
   (B) 75
   (C) 120
   (D) 150
   (E) 750

2. A company consists of two departments: Sales and production. If 30 percent of the 150 employees in the sales department received a holiday bonus and 70 percent of the 250 employees in the production department received a holiday bonus, what percent of all employees did not receive a holiday bonus?

(A) 40%
(B) 45%
(C) 50%
(D) 55%
(E) 60%

3. What is the ratio of $a$ to $b$?

(1) $a = b + 7$

(2) $3a = 4b$

(A) Statement (1) ALONE is sufficient, but statement (2) alone is not sufficient.
(B) Statement (2) ALONE is sufficient, but statement (1) alone is not sufficient.
(C) BOTH statements TOGETHER are sufficient, but NEITHER statement ALONE is sufficient.
(D) EACH statement ALONE is sufficient.
(E) Statements (1) and (2) TOGETHER are NOT sufficient.

4. All working at the same constant rate, 8 bartenders can pour 96 shots per minute. At this rate, how many shots could 3 bartenders pour in 2 minutes?

(A) 12
(B) 24
(C) 36
(D) 48
(E) 72

5. The average (arithmetic mean) of 2, 4, 6, and 8 equals the average of 1, 3, 5, and

(A) 7
(B) 9
(C) 10
(D) 11
(E) 12

6. The latest model of space shuttle can achieve a maximum speed of 25 miles per second. This maximum speed is how many miles per hour?

   (A) 1,500
   (B) 3,600
   (C) 9,000
   (D) 15,000
   (E) 90,000

7. How many of the employees at Company X have life insurance?

   (1) There are 300 employees at Company X.
   (2) The ratio of employees with life insurance to employees without life insurance is 1 : 5.

   (A) Statement (1) ALONE is sufficient, but statement (2) alone is not sufficient.
   (B) Statement (2) ALONE is sufficient, but statement (1) alone is not sufficient.
   (C) BOTH statements TOGETHER are sufficient, but NEITHER statement ALONE is sufficient.
   (D) EACH statement ALONE is sufficient.
   (E) Statements (1) and (2) TOGETHER are NOT sufficient.

8. What is $x$?

   (1) The ratio of $x : y$ is 1 : 3.
   (2) The ratio of $y : z$ is 1 : 2.

   (A) Statement (1) ALONE is sufficient, but statement (2) alone is not sufficient.
   (B) Statement (2) ALONE is sufficient, but statement (1) alone is not sufficient.
   (C) BOTH statements TOGETHER are sufficient, but NEITHER statement ALONE is sufficient.
   (D) EACH statement ALONE is sufficient.
   (E) Statements (1) and (2) TOGETHER are NOT sufficient.

9. The five starting players on a basketball team score points in the ratio of 1 : 1 : 2 : 3 : 4. If the starters score a total of 77 points in a particular game, how many points did the highest scoring starter score?

   (A) 7
   (B) 11
   (C) 14
   (D) 21
   (E) 28

10. If $1 were invested at 4 % interest, compounded quarterly, the total value of the investment, in dollars, at the end of 3 years would be

(A) $(1.4)^3$
(B) $(1.04)^{12}$
(C) $(1.04)^3$
(D) $(1.01)^{12}$
(E) $(1.01)^3$

11. Among a group of teenagers taking a driving test, 40 percent took a driver's education course. If 70 percent of the teenagers pass the driving test and all of those who took a driver's education course passed the test, what percent of the teenagers who did not take a driver's education course failed the test?

(A) 0%
(B) 30%
(C) 40%
(D) 50%
(E) 60%

12. If a laser printer can print 2 pages in 10 seconds, how many pages can it print in 3 minutes at the same rate?

(A) 5
(B) 12
(C) 18
(D) 36
(E) 60

13. If the average (arithmetic mean) of 3 positive integers is 35, how many of the numbers are greater than 10?

(1) The sum of 2 of the numbers is 75.
(2) None of the numbers is greater than 40.

(A) Statement (1) ALONE is sufficient, but statement (2) alone is not sufficient.
(B) Statement (2) ALONE is sufficient, but statement (1) alone is not sufficient.
(C) BOTH statements TOGETHER are sufficient, but NEITHER statement ALONE is sufficient.
(D) EACH statement ALONE is sufficient.
(E) Statements (1) and (2) TOGETHER are NOT sufficient.

14. Greg is training for a marathon by running to and from work each day, a distance of 12 miles each way. He runs from home to work at an average speed of 6 miles per hour and returns at an average speed of 4 miles per hour. What is Greg's average speed, in miles per hour, for the round trip?

    (A) 5.5
    (B) 5.0
    (C) 4.8
    (D) 2.5
    (E) 2.4

15. Each baseball team in a league has a roster of players in the ratio of 2 pitchers for every 3 fielders. If each team has a total of 25 players on its roster and there 12 teams in the league, the number of pitchers in the league is how much less than the number of fielders in the league?

    (A) 10
    (B) 15
    (C) 60
    (D) 120
    (E) 180

16. The first-year MBA class at XYZ University includes 60 male students and 40 female students. How many male students from the first-year class are taking economics?

    (1) Fifty percent of the first-year class is taking economics.
    (2) 10 female students from the first-year class are not taking economics.

    (A) Statement (1) ALONE is sufficient, but statement (2) alone is not sufficient.
    (B) Statement (2) ALONE is sufficient, but statement (1) alone is not sufficient.
    (C) BOTH statements TOGETHER are sufficient, but NEITHER statement ALONE is sufficient.
    (D) EACH statement ALONE is sufficient.
    (E) Statements (1) and (2) TOGETHER are NOT sufficient.

17. If the current population of country Z is 200,000 people and the population grows by 3% every year, which of the following expresses the population of Country Z in 5 years?

    (A) $200{,}000 \times (1.03)^5$
    (B) $200{,}000 \times 5 \times (1.03)$
    (C) $200{,}000 \times (0.03)^5$
    (D) $200{,}000 \times (1.3)^5$
    (E) $200{,}000 \times (1.05)^3$

18. The 250 students enrolled at ABC University are taking undergraduate courses, graduate courses, or both. How many students are taking graduate courses?

    (1) 200 students are taking undergraduate courses.

    (2) 50 students are taking both undergraduate and graduate courses.

    (A) Statement (1) ALONE is sufficient, but statement (2) alone is not sufficient.
    (B) Statement (2) ALONE is sufficient, but statement (1) alone is not sufficient.
    (C) BOTH statements TOGETHER are sufficient, but NEITHER statement ALONE is sufficient.
    (D) EACH statement ALONE is sufficient.
    (E) Statements (1) and (2) TOGETHER are NOT sufficient.

19. In a calculus class of 80 students, the ratio of math majors to non-math majors is 3 to 5. If 10 of the non-math majors drop out of the class, what fraction of the remaining students are non-math majors?

    (A) $\frac{3}{8}$

    (B) $\frac{3}{7}$

    (C) $\frac{4}{7}$

    (D) $\frac{5}{8}$

    (E) $\frac{3}{4}$

20. A total of 120 investment advisors work at a particular financial services firm, 30 in bonds and the rest in equities. Fifty percent of the investment advisors are board-certified. If one-third of the equities advisors are board-certified, how many bonds advisors are not board-certified?

    (A) 0
    (B) 10
    (C) 15
    (D) 20
    (E) 30

21. How many people contributed to the Charity Y?

(1) The average contribution to Charity Y was $100.

(2) Charity Y collected a total of $47,000 in contributions.

(A) Statement (1) ALONE is sufficient, but statement (2) alone is not sufficient.
(B) Statement (2) ALONE is sufficient, but statement (1) alone is not sufficient.
(C) BOTH statements TOGETHER are sufficient, but NEITHER statement ALONE is sufficient.
(D) EACH statement ALONE is sufficient.
(E) Statements (1) and (2) TOGETHER are NOT sufficient.

22. The average of two numbers is 108. What is the value of the greater number?

(1) The lesser number is 72.

(2) The ratio of the two numbers is 1 : 2.

(A) Statement (1) ALONE is sufficient, but statement (2) alone is not sufficient.
(B) Statement (2) ALONE is sufficient, but statement (1) alone is not sufficient.
(C) BOTH statements TOGETHER are sufficient, but NEITHER statement ALONE is sufficient.
(D) EACH statement ALONE is sufficient.
(E) Statements (1) and (2) TOGETHER are NOT sufficient.

## PROBLEM SET—ANSWERS AND EXPLANATIONS

1. (B) is correct. Set up a ratio box as shown below. The ratio total is $3 + 5 + 7 = 15$. Sharon's actual value is 15 and her ratio value is 3, so the multiplier is $15 \div 3 = 5$. Multiply all of the ratio values by 5 to get the actual values. So the actual total is $15 \times 5 = 75$. Choose (B).

| | Ratio | Multiplier | Actual |
|---|---|---|---|
| Sharon | **3** | | **15** |
| Carol | **5** | × 5 | 25 |
| Elinor | **7** | | 35 |
| Total | 15 | | **75** |

2. (B) is correct. Set up a grid as shown below. The number of sales employees receiving a bonus is 30% × 150 = 45, so the number of sales employees not receiving a bonus is 150 – 45 = 105. The number of production employees receiving a bonus is 70% × 250 = 175, so the number of production employees not receiving a bonus is 250 – 175 = 75. So 105 + 75 = 180 employees did not receive a bonus. The percent of all employees not receiving a bonus is $\frac{180}{150+250}=\frac{180}{400}=\frac{45}{100}=45\%$. Choose (B).

| | Sales | Production | Total |
|---|---|---|---|
| Bonus | 45 | 175 | 220 |
| No bonus | 105 | 75 | 180 |
| Total | **150** | **250** | 400 |

3. (B) is correct. Start with statement (1). You can't determine the ratio of $a$ to $b$. If $a = 8$ and $b = 1$, the ratio of $a$ to $b$ is 8 : 1. If $a = 9$ and $b = 2$, the ratio is 9 : 2. You can't answer the question, so eliminate (A) and (D). Try statement (2). You can rewrite this equation as $\frac{a}{b}=\frac{3}{x}$, so the ratio of $a$ to $b$ is 4 : 3. You can answer the question, so choose (B).
4. (E) is correct. You can set up equivalent fractions for this proportion: $\frac{8}{96}=\frac{3}{x}$. Cross-multiply to get $8x = 3 \times 96$ or $8x = 288$. Solving the equation, you get x = 36. However, this is just the number of shots 3 bartenders can pour in 1 minute. So in 2 minutes, 3 bartenders can pour 2 × 36 = 72 shots. Choose (E).
5. (D) is correct. Set up 2 average circles. In the first circle, the total is 2 + 4 + 6 + 8 = 20. The number of things is 4, so the average is $20 \div 4 = 5$. In the second circle the average is 5 (the same as the other one) and the number of things is 4. So the total should be 4 × 5 = 20. The missing number is 20 – 1 – 3 – 5 = 11. Choose (D).
6. (E) is correct. Set up the proportion: $\frac{25}{1}=\frac{x}{60\times 60}$. You use 60 × 60 for the number of seconds because there are 60 seconds in a minute and 60 minutes in an hour. Cross-multiply to get $x = 25 \times 60 \times 60$ or $x = 90{,}000$. Choose (E).

7. (C) is correct. Start with statement (1). This only tells you the total number of employees, not anything about the number that have life insurance. You can't answer the question, so eliminate (A) and (D). Look at statement (2). A ratio doesn't tell you anything about actual numbers unless you have another actual number to go with it (so you can find the multiplier). You can't answer the question, so eliminate (B). Try statements (1) and (2) together. Now you have the ratio as well as an actual number. You can use the actual number to find the multiplier, then use the multiplier to find all the other actual values. You can answer the question, so choose (C).

8. (E) is correct. Look at statement (1). The ratio won't alone won't help you find the actual values of the variables. You can't answer the question, so eliminate (A) and (D). Try statement (2). This doesn't even mention $x$, so you can't answer the question. Eliminate (B). Try statements (1) and (2) together. You could set up a ratio for $x$ to $z$. But you can't find the actual value of $x$ without another actual value to go with the ratio. You can't answer the question, so choose (E).

9. (E) is correct. Set up a ratio box as shown below. The ratio total is $1+1+2+3+4=11$. The actual total is 77, so the multiplier is $77 \div 11 = 7$. Multiply all of the ratio values by 7 to find the actual values. The actual value for the highest scoring starter is $4 \times 7 = 28$. Choose (E).

| | Ratio | Multiplier | Actual |
|---|---|---|---|
| Starter #1 | **1** | | 7 |
| Starter #2 | **1** | | 7 |
| Starter #3 | **2** | × 7 | 14 |
| Starter #4 | **3** | | 21 |
| Starter #5 | **4** | | 28 |
| Total | 11 | | **77** |

10. (D) is correct. You just need to insert the numbers into the compound interest formula: $(1+r)^t$. Interest compounds four times per year at a 4% rate, so $r = \frac{4\%}{4} = 1\% = 0.01$. The interest compounds 4 times per year for 3 years, so there are 12 interest periods. That means $t = 12$. The completed formula is $(1.01)^{12}$. Choose (D).

11. (D) is correct. Set up a grid as shown below. Plug in the numbers provided in the question (bold type in the diagram below). Note that you don't have actual numbers, so just use 100% for the total and the appropriate percentages for the other boxes. Start finding the other numbers. 40% took the course, so 60% didn't. 70% passed the test, so 30% didn't. All of the course takers (40%) passed, so 0% took the course and failed. A total of 70% passed, so 70% – 40% = 30% of teenagers passed and didn't take a course. That leaves 60% – 30% = 30% who did not take the course and failed. Out of the 60% of teenagers who didn't take the course, 30% of teenagers failed, but that's $\frac{30\%}{60\%} = 50\%$ of those who didn't take the course. Choose (D).

| | Course | No course | Total |
|---|---|---|---|
| Passed | **40%** | 30% | **70%** |
| Failed | 0% | 30% | 30% |
| Total | **40%** | 60% | **100%** |

12. (D) is correct. Set up the proportion: $\frac{2}{10} = \frac{x}{180}$. You need to keep the units consistent in both fractions, so use 180 seconds for the 3 minutes. Cross-multiply to get $10x = 2 \times 180$. That's $10x = 360$, so $x = 36$. Choose (D).

13. (B) is correct. Start with statement (1). The sum of the three numbers is $3 \times 35 = 105$. You can calculate that the third number is 105 – 75 = 30, definitely more than 10. However, you cannot tell whether 1 or 2 of the other numbers is over 10. They might be 40 and 35, or they might be 1 and 74. You can't answer the question, so eliminate (A) and (D). Try statement (2). This means that the smallest any number could be is 105 – 40 – 40 = 25. If you try to make a number smaller than 25, at least one of the numbers will be greater than 40. All 3 numbers must be more than 10. You can answer the question, so choose (B).

14. (C) is correct. On the way to work, Greg needs $\frac{12}{6} = 2$ hours to complete the trip. On the way home, Greg needs $\frac{12}{4} = 3$ hours to complete the trip. So he travels 12 + 12 = 24 miles in 2 + 3 hours. That's an average speed of $\frac{24}{5} = 4.8$ miles per hour. Choose (C). If you chose (B), you fell for the Joe Bloggs trap of averaging the averages.

15. (C) is correct. You need to use an expanded ratio box, like the one shown below. The ratio total is 2 + 3 = 5 players. There are 25 players on a team. So the multiplier for ratio to team is $25 \div 5 = 5$. Multiply all of the ratio values by 5 to find the actual team values. There are 12 teams in the league, so the multiplier for team to league is 12. Multiply all of the team values by 12 to get the league values. The number of fielders in the league is $3 \times 5 \times 12 = 180$. The number of pitchers in the league is $2 \times 5 \times 12 = 120$. So the difference is 180 – 120 = 60. Choose (C).

| | Ratio | Multiplier | Team | Multiplier | League |
|---|---|---|---|---|---|
| Pitchers | **2** | | 10 | | 120 |
| Fielders | **3** | × 5 | 15 | **× 12** | 180 |
| Total | 5 | | **25** | | 300 |

16. (C) is correct. You should set up a group grid as shown below. Start with statement (1). This allows you fill in 50 students for economics and 50 students not for economics. But you don't have two numbers anywhere that you can use to find a third number. You can't answer the question, so eliminate (A) and (D). Try statement (2). Erase any numbers you filled in from statement (1). Now you can find that 40 – 10 = 30 female students are taking economics. But you can't find how many male students are or are not taking economics because you don't have 2 of the 3 numbers. You can't answer the question, so eliminate (B). Try statements (1) and (2) together. Now you can find all the numbers because you have 2 out of every 3 numbers. Calculate all the numbers if you're unsure; otherwise, choose (C).

| | Course | No course | Total |
|---|---|---|---|
| Passed | **40%** | 30% | **70%** |
| Failed | 0% | 30% | 30% |
| Total | **40%** | 60% | **100%** |

17. (A) is correct. This is really an interest rate problem. The setup is exactly the same except that it's population growth instead of money growth. $r$ = 3% or 0.03 and $t$ = 5. So the completed formula is $200{,}000 \times (1.03)^5$. Choose (A).

18. (C) is correct. Check statement (1). From the group formula, you can set up 250 = 200 + Grad – Both + 0, but you can't solve the equation because there are 2 variables. You can't answer the question, so eliminate (A) and (D). Try statement (2). Now you have 250 = Undergrad + Grad – 50 + 0, but you still can't solve the equation. Eliminate (B). Try statements (1) and (2) together. Now you can fill in 250 = 200 + Grad – 50 + 0. You can solve for Grad and answer the question. Choose (C).

19. (C) is correct. Set up a ratio box as shown below. The ratio total is 8 and the actual total is 80, so the multiplier is 10. Multiply all the ratio values by 10 to get the actual values. The number of non-math majors is $5 \times 10 = 50$. 10 non-math majors drop out, leaving 40. That also reduces the total number of students from 80 to 70. So $\frac{40}{70} = \frac{4}{7}$ of the remaining students are non-math majors. Choose (C).

| | Ratio | Multiplier | Actual |
|---|---|---|---|
| Math | **3** | | 30 |
| Non-math | **5** | × 10 | 50 |
| Total | 8 | | **80** |

20. (A) is correct. Set up a group grid as shown below, with the given information in bold. If 30 of the 120 advisors work in bonds, the other 90 work in equities. The number of board-certified advisors is $50\% \times 120 = 60$, so the number of noncertified advisors is 120 – 60 = 60. The number of equity advisors that are certified is $\frac{1}{3} \times 90 = 30$, so the number of noncertified equities advisors is 90 – 30 = 60. There are 60 certified advisors and 30 of them work in equities, so 60 – 30 = 30 certified advisors work in bonds. So all 30 bond advisors are certified and none of them are noncertified. Choose (A).

| | Bonds | Equities | Total |
|---|---|---|---|
| Certified | 30 | 30 | 60 |
| Not certified | 0 | 60 | 60 |
| Total | **30** | 90 | **120** |

21. (C) is correct. Start with statement (1). There are 3 parts to any average: The total, the number of things, and the average. You know the average contribution, but you don't know the total. So you can't find the number of things (people). You can't answer the question. Eliminate (A) and (D). Try statement (2). Now you know the total contributions, but you don't know the average. You can't answer the question, so eliminate (B). Try statements (1) and (2) together. You know the average and the total, so you can find the number of things. You can answer the question, so choose (C).

22. (D) is correct. Start with statement (1). If the average of two things is 108, the total is $2 \times 108 = 216$. Now you can subtract 72 to find the other number. You can answer the question, so eliminate (B), (C), and (E). Try statement (2). You could set up a ratio box. You have the ratio and one actual number, so you could find all of the actual numbers. You can answer the question, so choose (D).

# 7

# Algebra Review

Many of the GMAT Math problems will involve algebra in some manner. Algebra is essentially arithmetic with one or more variables thrown into the mix. A variable is an unknown number and is usually indicated by an italicized letter, such as $x$ or $n$. In chapter 8, you'll learn some methods for avoiding much of the pain and many of the mistakes associated with algebra. However, there are times when you have no recourse but to solve the problem with algebra.

## SOLVING EQUATIONS

The golden rule for solving equations is: "Do the same thing to both sides until you isolate the variable." In other words you will add, subtract, multiply, and divide both sides by whatever means necessary to get the variable all by itself on one side of the equal sign. You should usually (but not always) add and subtract before you multiply and divide.

For example, solve for $x$ in the equation $3x + 7 = 19$.

| | |
|---|---|
| Subtract 7 from both sides to get | $3x = 12$ |
| Divide both sides by 3 to get | $x = 4$ |

You should also know how to deal with numbers in parentheses, such as $5(x + 3)$. In this case, just multiply each number in the parentheses by 5. So $5(x + 3) = 5x + 15$. You may also need to reverse this process by factoring. Just find the common factor of the numbers and put that outside of the parentheses. For example, factor $6x + 12$. Both numbers are divisible by 6, so factor out the 6 and put it in front of parentheses. So $6x + 12 = 6(x + 2)$.

Factoring is often required in problems with complicated fractions, such as $\frac{7x+14}{x+2}$. In order to simplify the fraction, you could try factoring out a number so that something on the top cancels with something on the bottom. This could let you reduce the fraction to a simpler form. If you factor out a 7 on the top to get $\frac{7x+14}{x+2}=\frac{7(x+2)}{x+2}$, then you can cancel the $x + 2$ on the top with the $x + 2$ on the bottom. So $\frac{7x+14}{x+2}=\frac{7}{1}=7$.

### QUIZ #1

1. If $\frac{1}{2}x+12=4$, then $x =$

   (A) –16
   (B) –8
   (C) –4
   (D) 8
   (E) 6

2. $\frac{5x+10}{x+2} =$

(A) 2
(B) 3
(C) 4
(D) 5
(E) 6

## Quiz #1 — Answers and Explanations

1. (A) is correct. Subtract 12 from both sides to get $\frac{1}{2}x = -8$. Then multiply both sides by 2 to get $x = -16$. Choose (A).
2. (D) is correct. First factor out 5 in the top of the fraction. That gives you $\frac{5(x+2)}{x+2}$. Now you can cancel the $x+2$ on the top with the one on the bottom to get $\frac{5}{1} = 5$. Choose (D).

## More Than One Variable (Simultaneous Equations)

You've seen how to solve an equation that contains one variable. What can you do if an equation has more than one variable, such as $2x + y = 7$? Well, you can't actually solve that equation without more information in the form of another equation. The general rule is that the number of different variables equals the number of different equations necessary to solve for those variables.

If you have two variables in two equations, such as $2x + y = 7$ and $x - y = -1$, you can solve for the variables. The method is often referred to as simultaneous equations. Line up the two equations as shown below. Then add the equations and eliminate one variable. This will leave you with an equation with one variable.

$$\begin{array}{r} 2x + y = 7 \\ + \; x - y = -1 \\ \hline 3x = 6 \end{array}$$

Solve that equation as you learned to do earlier in the chapter. $3x = 6$ becomes $x = 2$. Now you can find $y$ by plugging the value for $x$ into one of the original equations (it doesn't matter which). The first equation would become $2(2) + y = 7$. That gives you a one-variable equation, which you can then solve for $y$. In this case, $y = 3$.

What if simply adding the equations doesn't eliminate a variable, as with $x + y = 9$ and $2x + y = 12$. You need to multiply one or both of the equations to set up one of the variables for elimination. Multiply every number in the equation by the same number. In this example, you an multiply the first equation by $-1$. So $x + y = 9$ becomes $-x - y = -9$. Now you can add the equations and eliminate the $y$ variable. The resulting equation is $x = 3$. Plug that back into one of the equations. Using the first equation, you get $3 + y = 9$. Now you can solve for $y : y = 6$.

Look at a slightly more complex example: $3x + y = 11$ and $2x + 3y = 19$. In this case, you can multiply the first equation by –3 to eliminate the $y$ variable. Combining $-9x - 3y = -33$ with $2x + 3y = 19$, you get $-7x = -14$. So $x = 2$. Plug that into the first equation to get $3(2) + y = 11$ and solve to get $y = 5$.

## Quiz #2

1. If $x + 2y = 13$ and $5x - y = 21$, what is $x + y$ ?

   (A) 1
   (B) 4
   (C) 5
   (D) 8
   (E) 9

2. If $2a + 3b = 7$, what is the value of $b$ ?

   (1) $a - b = -4$

   (2) $4a = 14 - 6b$

   (A) Statement (1) ALONE is sufficient, but statement (2) alone is not sufficient to answer the question asked.
   (B) Statement (2) ALONE is sufficient, but statement (1) alone is not sufficient to answer the question asked.
   (C) BOTH statements (1) and (2) TOGETHER are sufficient to answer the question asked; but NEITHER statement ALONE is sufficient.
   (D) EACH statement ALONE is sufficient to answer the question asked.
   (E) Statements (1) and (2) TOGETHER are NOT sufficient to answer the question asked, and additional data specific to the problem are needed.

## Quiz #2—Answers and Explanations

1. (E) is correct. You can't eliminate a variable by adding the equations as they are. So multiply the second equation by 2 to eliminate the $y$ variable. Adding the equations, you get $11x = 55$. Divide both sides by 11 to get $x = 5$. Plug that back into the first equation to get $5 + 2y = 13$. Subtract 5 from each side to get $2y = 8$. Now divide by 2 to get $y = 4$. So $x + y = 5 + 4 = 9$. Choose (E).

2. (A) is correct. Start with statement (1). This gives you 2 different equations, so you can solve for the 2 variables. You can answer the question, so eliminate (B), (C), and (E). Look at statement (2). This does not give you 2 different equations. The equation from the statement is the same as the equation from the question. Take the question equation and multiply by 2

to get $4a + 6b = 14$. Then subtract $6b$ to get $4a = 14 - 6b$. If you tried to solve for the variables by adding the equations, you would find that everything was eliminated, leaving you with $0 = 0$. The key is that you need two *different* equations. You can't answer the question, so choose (A).

## INEQUALITIES

Inequalities are very similar to equations; instead of an equals sign (=), however, an inequality will use an inequality sign (>, <, ≥, or ≤). Those symbols mean "greater than," "less than," "greater than or equal to," and "less than or equal to," respectively. Solving an inequality is very much like solving an equality. You add, subtract, multiply, and divide to isolate the variable. For example, solve $2x + 4 > 6$.

| | |
|---|---|
| Subtract 4 to get | $2x > 2$ |
| Divide by 2 to get | $x > 1$ |

However, there is one critical difference between equalities and inequalities. When you multiply or divide an inequality by a negative number, you must reverse the inequality sign. For example, try $-3x + 4 < -2$.

| | |
|---|---|
| Subtract 4 to get | $4 - 3x < -6$ |
| Divide by –3 (reverse the inequality) to get | $x > 2$ |

Sometimes a problem will provide two inequalities and ask you to combine them in some way. Plug in the highest and lowest values for each variable and list all of the results. You'll have four numbers if there are two variables. From this list, choose the highest and lowest values and match the answer that includes those as the minimum and maximum values. Look at this next example:

1. If $2 < x < 5$ and $3 < y < 10$, which of the following expresses the possible range of values for $x - y$?

   (A) $-1 < x - y < 5$
   (B) $-1 < x - y < 2$
   (C) $-5 < x - y < 2$
   (D) $-8 < x - y < 2$
   (E) $-8 < x - y < -1$

Try all four combinations of the highest and lowest values for $x$ and $y$. If $x = 2$ (Well, obviously it couldn't be 2, because $2 < x$, but it could be very close, for instance 2.01. Here we'll use 2 as the closest approximation.) and $y = 3$, then $x - y = 2 - 3 = -1$. If $x = 2$ and $y = 10$, then $x - y = 2 - 10 = -8$. If $x = 5$ and $y = 3$, then $x - y = 2$. Finally, when $x = 5$ and $y = 10$, then $x - y = -5$. The highest value for $x - y$ is 2 and the lowest value is –8. Those are the numbers you want in the answer, so choose (D).

## Quiz #3

1. If $5 - 3x > 3$, which of the following expresses the possible values of $x$ ?

   (A) $x < -\frac{2}{3}$

   (B) $x > -\frac{2}{3}$

   (C) $x < \frac{2}{3}$

   (D) $x > \frac{2}{3}$

   (E) $x > \frac{3}{2}$

2. If $-2 < p < 3$ and $-1 < q < 10$, which of the following expresses the possible values of $p - q$ ?

   (A) $-12 < p - q < 4$
   (B) $-12 < p - q < 7$
   (C) $-3 < p - q < 4$
   (D) $-3 < p - q < 7$
   (E) $-1 < p - q < 4$

## Quiz #3—Answers and Explanations

1. (C) is correct. Subtract 5 from each side of the inequality to get $-3x > -2$. Then divide each side by $-3$ to get $x < -\frac{2}{3}$. Remember, you need to flip the inequality sign when you divide (or multiply) by a negative number. Choose (C).

2. (A) is correct. To combine the inequalities, calculate $p - q$ using all four numbers. When $p = -2$ and $q = -1$ then $p - q = -2 - (-1) = -2 + 1 = -1$. When $p = -2$ and $q = 10$, then $p - q = -2 - 10 = -12$. When $p = 3$ and $q = -1$, then $p - q = 3 - (-1) = 3 + 1 = 4$. Finally, when $p = 3$ and $q = 10$, then $p - q = 3 - 10 = -7$. The greatest value of $p - q$ is 4 and the least value is $-12$, so $-12 < p - q < 4$. Choose (A).

## EXPONENTS

Exponents are just a shorthand expression for multiplication. For example, $4^5$ simply means $4 \times 4 \times 4 \times 4 \times 4$. If you ever get confused by exponents, rewriting them as multiplication can demystify the problem.

When two exponents have the same base, you can combine them in several ways. Here are the rules for manipulating exponents:

| | |
|---|---|
| To multiply, add the exponents | $3^2 \times 3^5 = 3^{2+5} = 3^7$ |
| To divide, subtract the exponents | $\frac{7^5}{7^2} = 7^{5-2} = 7^3$ |
| To raise to a power, multiply the exponents | $\left(5^2\right)^3 = 5^{2\times 3} = 5^6$ |
| To find a root, divide the exponents | $\sqrt{5^8} = 5^{8\div 2} = 5^4$ |

Remember, you can also rewrite the exponents as multiplication if you forget these rules.

$$3^2 \times 3^5 = 3 \times 3 \times 3 \times 3 \times 3 \times 3 \times 3 = 3^7$$

$$\frac{7^5}{7^2} = \frac{7 \times 7 \times 7 \times 7 \times 7}{7 \times 7} = 7 \times 7 \times 7 = 7^3$$

$$\left(5^2\right)^3 = (5 \times 5)(5 \times 5)(5 \times 5) = 5^6$$

There a few special cases for exponents that you should know:

| | |
|---|---|
| Any number to the zero power equals 1. | $x^0 = 1$ |
| Any number to the first power equals itself. | $x^1 = x$ |
| One to any power equals 1. | $1^x = 1$ |

It's possible that you'll see a fractional exponent on the GMAT. They're really not that tough. A fractional exponent is just a way of expressing an exponent and a root at the same time. The top of the fraction is the exponent and the bottom part is the root. See the example below:

$$x^{\frac{3}{4}} = \sqrt[4]{x^3}$$

You may also see a negative exponent. The negative sign just means to divide by the positive exponent. For example:

$$x^{-3} = \frac{1}{x^3}$$

Scientific notation is a way to express very large and very small numbers (such as the number of inches between the earth and the sun) in a readable format. You move the decimal point so that there's one digit to the left of the decimal point and then you multiply by a power of 10. The exponent represents the number of places you moved the decimal point. A positive exponent means the original number was big (more than 1) and a negative exponent means the original number was small (less than 1). For example:

$$1{,}200{,}000 = 1.2 \times 10^6$$

$$0.0000567 = 5.67 \times 10^{-5}$$

## Quiz #4

1. $\frac{4^5 + 4^3}{4^2} =$

   (A) $4^2$

   (B) $4^{\frac{5}{2}} + 4^{\frac{3}{2}}$

   (C) $4^3 + 4$

   (D) $4^4$

   (E) $4^6$

2. If $x = -3$, then $\frac{x^3 - x^2 - x}{x + 1} =$

   (A) $-\frac{33}{2}$

   (B) $-\frac{39}{4}$

   (C) $\frac{33}{4}$

   (D) $\frac{33}{2}$

   (E) $\frac{39}{2}$

3. The number of people surfing the Internet doubles every 3 months. If the number of people surfing the Internet today is $10^3$, how many people will be surfing the Internet in 1 year?

   (A) $4 \times 10^3$
   (B) $10^3 \times 2^4$
   (C) $10^7$
   (D) $10^7 \times 2^4$
   (E) $10^{11}$

## Quiz #4—Answers and Explanations

1. (C) is correct. First factor out $4^2$ on the top of fraction. That gives you $\frac{4^2(4^3+4)}{4^2}$. Cancel the $4^2$ on the top with the one on the bottom. That gives you $4^3 + 4$. Choose (C).
2. (D) is correct. Plug $x = -3$ into the fraction to get: $\frac{-27-9-(-3)}{-3+1} = \frac{-27-9+3}{-2} = \frac{-33}{-2} = \frac{33}{2}$. Don't forget about the positive/negative rules for multiplication. They say that a negative number to an even exponent will be positive and a negative number to an odd exponent will be negative. Choose (D).
3. (B) is correct. Every time the number of Internet surfers doubles, you're just multiplying by 2. It will double four times in a year, so that's $2 \times 2 \times 2 \times 2$ or $2^4$. You need to multiply that by the original $10^3$ to get $10^3 \times 2^4$. Choose (B).

# ROOTS

A root is like an exponent in reverse. For example, $5^3 = 125$ and $\sqrt[3]{125} = 5$. So the square root of $x$ is the number that you would square to get $x$. Almost all of the roots on the GMAT are square roots (as opposed to cube roots and so forth). Square roots are generally indicated by a radical sign ($\sqrt{\ }$); other roots will include a small number over the left side of the radical sign that tells you what kind of root it is. For example, the 3 in $\sqrt[3]{125}$ tells you to find the third root (or cube root).

Adding and subtracting roots is jut like adding and subtracting variables. If the roots are the same, you can add and subtract. If they're different, you can't add or subtract. Just as you can add $x + 4x = 5x$, you can add $\sqrt{3} + 4\sqrt{3} = 5\sqrt{3}$. Just as you can't do anything with $x + y$, you can't combine $\sqrt{2} + \sqrt{3}$. You can only add or subtract if the roots are the same.

If you're multiplying or dividing several roots, it's usually easiest to combine them under one radical sign and then do the calculation. Look at these examples:

$$2\sqrt{2} \times \sqrt{8} = 2 \times \sqrt{2 \times 8} = 2\sqrt{16} = 2 \times 4 = 8$$

$$\frac{\sqrt{12}}{\sqrt{3}} = \sqrt{\frac{12}{3}} = \sqrt{4} = 2$$

Sometimes you need to take a single root and split it up into several roots. Try to factor out perfect squares so that you can simplify them to integers. Look at these examples:

$$\sqrt{\frac{1}{4}} = \frac{\sqrt{1}}{\sqrt{4}} = \frac{1}{2}$$

$$\sqrt{48} = \sqrt{16 \times 3} = \sqrt{16} \times \sqrt{3} = 4\sqrt{3}$$

Sometimes your answer will have a root in the bottom of a fraction. If it does, that answer won't be in the answer choices. On the GMAT you can't have a radical in the denominator of a fraction, so you'll need to convert it to something that's acceptable. Just multiply both the top and bottom of the fraction by that root. It will get rid of the root in the bottom. You haven't changed the value of the number, just what it looks like. For example:

$$\frac{3\sqrt{2}}{\sqrt{5}} = \frac{3\sqrt{2}}{\sqrt{5}} \times \frac{\sqrt{5}}{\sqrt{5}} = \frac{3\sqrt{10}}{5}$$

A number has both positive and negative roots. For example, the square root of 9 can be either 3 or –3. That's because $3^2 = 9$ and also $(-3)^2 = 9$. Most of the time, you'll be dealing with the positive root. On the GMAT, the radical sign refers only to the positive root. So, $x = \sqrt{9}$ means $x = 3$. But if $x^2 = 9$ then $x = 3$ or $-3$. It's an important distinction, especially on Data Sufficiency questions.

## QUIZ #5

1. $\sqrt{10} \times \sqrt{2} + \frac{\sqrt{10}}{\sqrt{2}}$

(A) $\sqrt{2} + \sqrt{5}$

(B) $2\sqrt{5}$

(C) $3\sqrt{5}$

(D) $4\sqrt{5}$

(E) $3\sqrt{10}$

2. What is $n$ ?

   (1) $n = \sqrt{25}$

   (2) $n^2 = 25$

   (A) Statement (1) ALONE is sufficient, but statement (2) alone is not sufficient to answer the question asked.
   (B) Statement (2) ALONE is sufficient, but statement (1) alone is not sufficient to answer the question asked.
   (C) BOTH statements (1) and (2) TOGETHER are sufficient to answer the question asked; but NEITHER statement ALONE is sufficient.
   (D) EACH statement ALONE is sufficient to answer the question asked.
   (E) Statements (1) and (2) TOGETHER are NOT sufficient to answer the question asked, and additional data specific to the problem are needed.

## Quiz #5—Answers and Explanations

1. (C) is correct. When you multiply and divide roots, you can combine them under one radical sign. But you can't do that with adding or subtracting roots. So you can combine $\sqrt{10} \times \sqrt{2} = \sqrt{10 \times 2} = \sqrt{20}$. You can also combine $\frac{\sqrt{10}}{\sqrt{2}} = \sqrt{\frac{10}{2}} = \sqrt{5}$. Putting this together, you get $\sqrt{20} + \sqrt{5}$. You can factor the perfect square 4 out of 20 to get $\sqrt{20} + \sqrt{5} = \sqrt{4 \times 5} + \sqrt{5} = \sqrt{4} \times \sqrt{5} + \sqrt{5} = 2\sqrt{5} + \sqrt{5}$. Since the base is the same you can add $2\sqrt{5} + \sqrt{5} = 3\sqrt{5}$. Choose (C).
2. (A) is correct. Start with statement (1). You know that the radical sign refers only to the positive root. So $n = 5$. You can answer the question. Eliminate (B), (C), and (E). Try statement (2). In this case, $n$ could be either the positive root (5) or the negative root (–5). You have more than one value, so you can't answer the question. Choose (A).

# QUADRATIC EQUATIONS

On the GMAT, you will most likely see some problems that involve quadratic equations and polynomials. The good news is you don't really have to know what either of those terms mean (or even how to spell them). Just follow a few guidelines to get you through these problems.

One issue that will probably come up is multiplying two sets of parentheses, such as $(x + 2)(x + 3)$. To do this, remember FOIL. It stands for First, Outer, Inner, Last. Those are the pairs of numbers you need to multiply. Check out the diagram below:

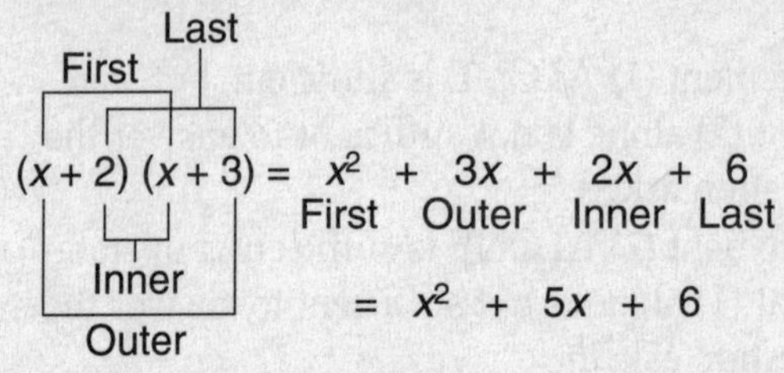

Last

First

$$(x + 2)\ (x + 3) = x^2 + 3x + 2x + 6$$

First Outer Inner Last

Inner

$$= x^2 + 5x + 6$$

Outer

Notice that you can usually combine the Inner and Outer terms into one term.

Sometimes you'll be given an equation that you need to factor out. Start by looking at the first number. That will tell you the first number in each set of parentheses. Then look at the signs (positive or negative) of the middle number and last number of the original equation. These tell you the signs of the second numbers inside the parentheses. Then find the factors of that last number because those are the potential choices for the second numbers in each set of parentheses. Last, look at the middle number of the original equation to help choose which pair of factors is appropriate. Choose the numbers that add up to this middle number. Be very careful with positive and negative signs.

Suppose you need to factor $x^2 - 4x + 4$. From the $x^2$, you can tell that the first numbers in the parentheses are $x$ and $x$. From the sign of the last number (+4), you can tell that you need two positives or two negatives because the product must be positive. The negative sign of $-4x$ tells you it must be two negatives. The factors of the 4 are $-1$ and $-4$ or $-2$ and $-2$ (you already know that they're negative). To add up to $-4x$, you need $-2$ and $-2$. So the equation $x^2 - 4x + 4$ factors out to $(x - 2)(x - 2)$.

$$\begin{aligned} x^2 - 4x + 4 &= (x \quad)(x \quad) \\ &= (x - \quad)(x - \quad) \\ &= (x - 2)(x - 2) \end{aligned}$$

There are three patterns of quadratic equations that the GMAT writers love to use. You should memorize these and look for them whenever you see a problem with quadratic equations. You'll be able to avoid the tedious calculations if you recognize the patterns. They are:

$$(x + y)(x + y) = x^2 + 2xy + y^2$$
$$(x - y)(x - y) = x^2 - 2xy + y^2$$
$$(x + y)(x - y) = x^2 - y^2$$

The final thing you need to know about quadratic equations is how to use them to solve for a variable. First move everything to one side to get a zero on the other side. Factor out the non-zero side. Set each of the parentheses equal to zero. You do that because if the product of two numbers is zero, one of the numbers must be zero. Solve both of these new mini-equations to get the value of the variable. Note that this gives you two answers (unless the two sets of parentheses are the same). However, this is the best you can do with a quadratic equation. These answers are often referred to as the solutions or "roots" of the equation, so don't let those terms confuse you.

Suppose you have the equation $x^2 - 4x + 7 = 4$ and you need to solve for $x$. First subtract 4 from both sides to get a zero on one side. The equation is now $x^2 - 4x + 3 = 0$. Now factor the equation into $(x - 1)(x - 3)$ using the guidelines you learned earlier. Last, set up and solve the two mini-equations: $x - 1 = 0$ and $x - 3 = 0$. The solutions are $x = 1$ or $x = 3$.

## Quiz #6

1. If $\frac{x^2+5x+6}{x+2} = 5$, what is the value of $x + 5$?

   (A) 2
   (B) 3
   (C) 5
   (D) 7
   (E) 10

2. If $x^2 - 5x - 6 = 0$, which of the following could be $x$ ?

   (A) −2
   (B) −1
   (C) 1
   (D) 2
   (E) 3

## Quiz #6—Answers and Explanations

1. (D) is correct. First factor the top of the fraction. The $x^2$ tells that you have $x$ and $x$ for the first numbers inside the parentheses. The signs of the middle ($+ 5x$) and last numbers ($+ 6$) tell you that you're adding in both of the parentheses. The factors of 6 are 1 and 6 or 2 and 3. To add up to 5, you need 2 and 3. So $x^2 + 5x + 6$ factors into $(x + 2)(x + 3)$ to give you $\frac{(x+2)(x+3)}{x+2} = 5$. You can cancel the $x + 2$ in the top and the bottom of the fraction to get $x + 3 = 5$, so $x = 2$. However, the question asks for $x + 5$. So $x + 5 = 2 + 5 = 7$. Choose (D).

2. (B) is correct. Factor the quadratic equation. The $x^2$ tells you that the first numbers in the parentheses are $x$ and $x$. The signs of the middle and last numbers tell you that you will add in one parentheses and subtract in the other. The factors of 6 are 1 and 6 or 2 and 3. Since one is positive and one is negative, and since they must add to −5, you must use −6 and 1. So the equation becomes $(x - 6)(x + 1) = 0$. Set up the two mini-equations $x - 6 = 0$ and $x + 1 = 0$. Solving them, you get $x = 6$ or $x = -1$. The only answer that matches is (B).

## PROBLEM SET

1. Which one of the following equations has a root in common with $m^2 + 4m + 4 = 0$?

   (A) $m^2 - 4m + 4 = 0$
   (B) $m^2 + 4m + 3 = 0$
   (C) $m^2 - 4 = 0$
   (D) $m^2 + m - 6 = 0$
   (E) $m^2 + 5m + 4 = 0$

2. If $\frac{3}{1+\frac{3}{x}} = 2$, then $x =$

   (A) $-6$
   (B) $-\frac{3}{2}$
   (C) $\frac{1}{2}$
   (D) $\frac{3}{2}$
   (E) $6$

3. If $3^{n+1} = 9^{n-1}$, then $n =$

   (A) $-1$
   (B) $0$
   (C) $1$
   (D) $2$
   (E) $3$

4. If $7 < m < 11$ and $-2 < n < 5$, then which of the following expresses the possible values of $mn$?

   (A) $-14 < mn < 35$
   (B) $-14 < mn < 55$
   (C) $-22 < mn < 35$
   (D) $-22 < mn < 55$
   (E) $5 < mn < 16$

5. Which of the following most closely approximates $\sqrt{\frac{(3.97)(50.02)}{1.84}}$?

(A) 1
(B) 5
(C) 10
(D) 20
(E) 100

6. If $x + 3y = 15$ and $y - x = 5$, then $y =$

(A) –5
(B) 0
(C) 3
(D) 5
(E) 15

7. If $n = \frac{n-6}{n+6}$, what is the value of $n^2 + 5n + 6$?

(A) –3
(B) –2
(C) 0
(D) 2
(E) 3

8. If $\frac{2.25 \times 10^a}{0.15 \times 10^b} = 1.5 \times 10^3$, then $a - b =$

(A) 1
(B) 2
(C) 3
(D) 4
(E) 5

9. $(3 + \sqrt{11})(3 - \sqrt{11})$

(A) –2
(B) $6 - 2\sqrt{11}$
(C) 2
(D) $9 - 2\sqrt{11}$
(E) 8

10. If $(x - 2)^2 = 100$, which of the following could be the value of $x + 2$?

(A) $-10$
(B) $-8$
(C) $-6$
(D) 10
(E) 12

11. If 3 is one value of $x$ for the equation $x^2 - 7x + k = -5$, where $k$ is a constant, what is the other solution?

(A) 2
(B) 4
(C) 5
(D) 6
(E) 12

12. $\dfrac{19^2 + 19}{20}$

(A) $\dfrac{19}{10}$
(B) $\dfrac{57}{20}$
(C) 19
(D) 38
(E) 361

13. $\dfrac{(0.2)^2}{(0.2)^4}$

(A) 0.04
(B) 0.25
(C) 0.4
(D) 4.0
(E) 25.0

14. $\sqrt{25 + 25}$

(A) $2\sqrt{5}$
(B) $5\sqrt{2}$
(C) 10
(D) 20
(E) 25

15. If $3x - 5 = 15 - 2x$, then $2x =$

(A) 2
(B) 4
(C) 6
(D) 8
(E) 10

## Problem Set—Answers and Explanations

1. (C) is correct. Find the factors of the equation in the question and then factor the answer choices until you find one that matches. The equation in the question factors to $(m+2)(m+2) = 0$. Answer (A) factors to $(m-2)(m-2) = 0$. Answer (B) factors to $(m + 3)(m + 1) = 0$. Answer (C) factors to $(m + 2)(m - 2) = 0$. Answer (D) factors to $(m + 3)(m - 2) = 0$. Answer (E) factors to $(m + 4)(m + 1) = 0$. The only one that matches is the $(m + 2)$ from (C).

2. (E) is correct. First multiply both sides by $1+\frac{3}{x}$. This gives you $3=2+\frac{6}{x}$. Next multiply both sides by $x$ to get $3x = 2x + 6$. Subtract $2x$ from both sides to get $x = 6$. Choose (E).

3. (E) is correct. You can't combine exponents unless the bases are the same. So translate the 9 into $3^2$. That gives you $3^{n+1} = \left(3^2\right)^{n-1}$. To raise an exponent to a power, just multiply the exponents. That gives you $3n + 1 = 3^2n - 2$. So $n + 1 = 2n - 2$. Add 2 to both sides to get $n + 3 = 2n$. Subtract $n$ from each side to get $3 = n$. Choose (E).

4. (D) is correct. Plug in both of the values for $m$ and both of the values for $n$ to get the four possible numbers. They are –14, 35, –22, and 55. The greatest value is 55 and the least value is –22, so $-22 < mn < 55$. Choose (D).

5. (C) is correct. The problem says "approximates," so round off the numbers and Ballpark. The problem becomes $\sqrt{\frac{4\times 50}{2}} = \sqrt{\frac{200}{2}} = \sqrt{100} = 10$. Choose (C).

6. (D) is correct. Rewrite the second equation as $-x + y = 5$, then add the equations. The $x$ variable is eliminated, giving you $4y = 20$. That means that $y = 5$. You don't even need to plug $y$ back in to find $x$, unless you just can't stand not knowing. Choose (D).

7. (C) is correct. Multiply both sides of the equation by $n + 6$ to get $n^2 + 6n = n - 6$. Subtract $n$ and add 6 to both sides to get $n^2 + 5n + 6 = 0$. Surprisingly enough, that's the value you need. Choose (C).

8. (B) is correct. Split the decimals and the powers of 10 into separate fractions to get $\frac{2.25}{0.15}\times\frac{10^a}{10^b}=1.5\times10^3$. When you do the division, you get $15\times\frac{10^a}{10^b}=1.5\times10^3$. When you divide numbers with exponents, just subtract the exponents. So the equation becomes $15\times10^{a-b}=1.5\times10^3$. Divide both sides by 15 to get $10^{a-b}=0.1\times10^3$, which becomes $10^{a-b}=10^2$. So $a-b=2$. Choose (B).

9. (A) is correct. Use FOIL to multiply these terms. You get $9-3\sqrt{11}+3\sqrt{11}-11=9-11=-2$. Notice that this is one of the three common patterns for quadratic equations. Choose (A).

10. (C) is correct. If you take the square root of both sides, you get $x-2=10$ or $x-2=-10$. Remember that a number has both positive and negative roots. Solving both of these equations, you get $x=12$ and $x=-8$. However, the question asks for $x+2$, which will be $12+2=14$ or $-8+2=-6$. The only answer that matches is (C).

11. (B) is correct. The question tells you that $x$ can equal 3. So plug that into the equation to solve for $k$. You get $3^2-7(3)+k=-5$ which becomes $9-21+k=-5$. So $-12+k=-5$ or $k=7$. Now you can plug that into the original equation to solve for the other possible value of $x$. You get $x^2-7x+7=-5$. Add 5 to each side to get $x^2-7x+12=0$. That factors into $(x-4)(x-3)=0$. Set up the two mini-equations $x-4=0$ and $x-3=0$. Solving them, you get $x=4$ or $x=3$. So the two possible values for $x$ are 4 (the new one) and 3 (which you already had). Choose (B).

12. (C) is correct. Factor out a 19 in the top of the fraction to get $\frac{19(19+1)}{20}=\frac{19(20)}{20}=19$. Choose (C).

13. (E) is correct. You can cancel a $(0.2)^2$ from the top and bottom of the fraction, leaving you with $\frac{1}{(0.2)^2}=\frac{1}{0.04}=25$. Choose (E).

14. (B) is correct. Don't fall for the Joe Bloggs trap of 25. Add the numbers to get $\sqrt{25+25}=\sqrt{50}$. Then factor out the perfect square to get $\sqrt{50}=\sqrt{25\times2}=\sqrt{25}\times\sqrt{2}=5\sqrt{2}$. Choose (B).

15. (D) is correct. Add $2x$ to each side to get $5x-5=15$. Then add 5 to each side to get $5x=20$. Then divide by 5 to get $x=4$. However, the question asks for $2x$, so $2(4)=8$. Choose (D).

# 8

# Avoiding Algebra

As you saw in Chapter 7, algebra is no fun. There are plenty of seemingly arbitrary rules and it's very easy to make a mistake. In this chapter, you'll learn some powerful methods for bypassing the algebra in a lot of questions. You'll need some practice to be comfortable using these techniques and recognize opportunities to use them. First, however, a question:

Which of the following two problems would you rather see on the GMAT?

1. Otto has a five-dollar bill. He goes to the store and buys 3 pieces of candy that cost 50 cents each. How much change, in dollars, does Otto receive?

   (A) $0.50
   (B) $1.50
   (C) $2.00
   (D) $3.00
   (E) $3.50

2. Oscar has an $x$-dollar bill. He goes to the store and buys $y$ pieces of candy that cost $z$ cents each. How much change, in dollars, does Oscar receive?

   (A) $x - yz$
   (B) $xy - z$
   (C) $x - \frac{yz}{100}$
   (D) $100x - yz$
   (E) $\frac{x - yz}{100}$

Why does the second question seem so much harder? It's identical to the first question except that it contains variables instead of numbers. So it must be the variables that make it harder. In fact, algebra is always harder than arithmetic. To put it another way, no matter how good you are at algebra, you are certainly better at arithmetic. The techniques in this chapter will help you transform algebra questions into arithmetic questions and make the world a better place.

By the way, the answer to question #2 is <u>not</u> (A). You'll see how to solve it in a short while.

## PLUGGING IN

As you saw with the problem about Oscar's candy, variables make a question considerably tougher. Wouldn't it be wonderful if there were a way to get rid of those variables? Well, there is. It's a method called Plugging In. Whenever you see a problem with variables in the answer choices, you should consider this approach. Here's the method:

1. Replace every variable in the problem with a number. Just make one up.
2. Work the problem using the numbers you plugged in. You'll get a number for an answer.

3. Plug your made-up numbers into the variables in the answers. Check all five and see which one matches your answer from step 2.

Try this approach with the Oscar example:

2. Oscar has an $x$-dollar bill. He goes to the store and buys $y$ pieces of candy that cost $z$ cents each. How much change, in dollars, does Oscar receive?

(A) $x - yz$
(B) $xy - z$
(C) $x - \frac{yz}{100}$
(D) $100x - yz$
(E) $\frac{x - yz}{100}$

First, make up numbers for the variables. Maybe $x = 10$ dollars, $y = 4$ pieces of candy, and $z = 50$ cents. You can try a different set if you like. Plugging In works with any numbers.

Second, work the problem using these numbers. If Oscar buys 4 pieces of candy for 50 cents each, he spends $4 \times 50 = 200$ cents or 2 dollars. He started with 10 dollars, so he gets $10 - 2 = 8$ dollars in change. It's important to make sure that your answer is in the units (dollars or cents) specified by the problem. So the numerical answer is 8.

Third, plug your made-up numbers into the answer choices. Using $x = 10$, $y = 4$, and $z = 50$, the answers are:

(A) $x - yz = 10 - (4)(50) = 10 - 200 = -190$
(B) $xy - z = (10)(4) - 50 = 40 - 50 = -10$
(C) $x - \frac{yz}{100} = 10 - \frac{(4)(50)}{100} = 10 - \frac{200}{100} = 10 - 2 = 8$
(D) $100x - yz = (100)(10) - (4)(50) = 1{,}000 - 200 = 800$
(E) $\frac{x - yz}{100} = \frac{10 - (4)(50)}{100} = \frac{10 - 200}{100} = \frac{-190}{100} = -1.90$

As you can see, the answer must be (C).

You've seen how Plugging In works when the answer choices contain variables, but sometimes the variable in the answers isn't quite so obvious. In fact, it may be invisible. If the answers contain fractions or percents, check the question. If it asks for a fraction or percent of some unknown amount, that unknown amount is the invisible variable. Plug In a number for that variable and use the steps you just learned. Follow this example:

3. Marty spends $\frac{1}{3}$ of his weekly allowance on baseball cards.

He spends $\frac{1}{4}$ of the rest on bubble gum. He spends $\frac{1}{6}$ of his allowance on soda pop. If Marty has no other expenses and saves the rest of his allowance in a piggy bank, what fraction of his allowance does he save in the piggy bank?

(A) $\frac{1}{6}$

(B) $\frac{1}{4}$

(C) $\frac{1}{3}$

(D) $\frac{5}{12}$

(E) $\frac{1}{2}$

The question asks for a fraction of Marty's allowance, so Marty's allowance is the invisible variable in the answer choices. First, make up a number, say $60, for Marty's allowance. Second, work the problem using that number. Marty spends $\frac{1}{3} \times 60 = 20$ on baseball cards. That leaves 60 – 20 = 40 to spend. He spends $\frac{1}{4}$ *of the rest* or $\frac{1}{4} \times 40 = 10$ on bubble gum. That leaves 40 – 10 = 30 dollars. He spends $\frac{1}{6} \times 60 = 10$ on soda pop, leaving 30 – 10 = 20 dollars for his piggy bank. Third, you need to convert that number to match the answers. He saves 20 dollars out of the original 60 or $\frac{20}{60} = \frac{1}{3}$. The answer is (C).

Whether the variable is visible or invisible, you can make the Plugging In method easier to use by choosing good numbers. Although any number will work, some numbers make the calculations unnecessarily difficult. For example, you wouldn't want to use $11.17 for Marty's allowance in the previous example. That amount wouldn't work well with the other numbers in the problem: $\frac{1}{3}$, $\frac{1}{4}$, and $\frac{1}{6}$. Instead, you want to choose a number that is divisible by 3, 4, and 6.

In some cases, the particular numbers you choose can make more than one answer seem correct. For example, suppose you Plug In $x = 4$ and your eventual result is 16. If two of the answers are $4x$ and $x^2$, they will both seem to be correct. If that happens to you, just rework the problem with a different number. You only need to check the answers that seemed correct the first time.

Of course, it would be better if you could prevent that problem in the first place. One way to do that is to avoid using the exact numbers from the problem or the answer choices. Instead, use multiples of those numbers.

Use these guidelines to choose good numbers for this next problem:

4. A machine can produce $p$ fan blades in an hour. How many fan blades can this machine produce in $q$ minutes?

(A) $pq$

(B) $\frac{60p}{q}$

(C) $60pq$

(D) $\frac{p}{60q}$

(E) $\frac{pq}{60}$

First Plug In some numbers for the variables. Suppose $p = 10$ fan blades in an hour and $q = 60$ minutes.

Second, find a numerical answer. The machine produces 10 fan blades in an hour, so it produces 10 fan blades in 60 minutes, because 60 minutes is one hour. So the answer is 10.

Third, plug your numbers into the answer choices to see which one matches. For (A), you get $pq = (10)(60) = 600$. The answer should be 10, so eliminate (A). For (B), you get $\frac{60p}{q} = \frac{60 \times 10}{60} = 10$. That seems right. However, you remember that you're supposed to check all five choices, so you keep going. For (C), you get $60pq = (60)(10)(60) = 36{,}000$. That's certainly not 10, so eliminate (C). For (D), you get $\frac{p}{60q} = \frac{10}{60 \times 60} = \frac{10}{3{,}600} = \frac{1}{360}$. That's not 10, so eliminate (D). For (E), you get $\frac{pq}{60} = \frac{10 \times 60}{60} = 10$. Wait a second, (B) and (E) can't both be right! Now you need to try some different numbers and check (B) and (E) again.

Where did this all go wrong? The problem started with setting $q = 60$. Because 60 shows up in the answer choices, it's possible that $q = 60$ will make some answers equal in value. To avoid that problem, use a different value for $q$. You want to keep the calculations simple, so use a number that "plays nicely" with 60, such as 30 or 120. Start with these numbers so that you don't have to work the problem twice.

Try the problem with a different number for $q$, such as $q = 120$ minutes. Keep $p = 10$ fan blades per hour. In 120 minutes, the machine can produce 20 fan blades, because 120 minutes is 2 hours. So the answer is 20 fan blades. When you plug $p = 10$ and $q = 120$ into the answer choices, you get:

(A) $pq = (10)(120) = 1{,}200$

(B) $\frac{60p}{q} = \frac{60 \times 10}{120} = 50$

(C) $60pq = (60)(10)(120) = 72{,}000$

(D) $\frac{p}{60q} = \frac{10}{60 \times 120} = \frac{1}{7200}$

(E) $\frac{pq}{60} = \frac{10 \times 120}{60} = 20$

So the correct answer turns out to be (E).

There are also a couple of guidelines you should follow when using Plugging In on a problem containing several variables. First, use a different number for each variable. Failure to do so can result in several answers all seeming to work. This is essentially the same problem as using a number that appears in the answer choices.

Second, if the problem contains several variables all related in an equation, you're going to make up numbers for all the variables except one. Solve the equation to find the value for that last variable. Here's an example:

5. If $a = \frac{b+5}{c}$, what is the value of $b$ in terms of $a$ and $c$?

(A) $\frac{ac}{5}$

(B) $ac - 5$

(C) $c(a - 5)$

(D) $(a - 5)(c - 5)$

(E) $5ac$

First, Plug In numbers for the variables. There is an equation, so you'll make up numbers for only two of the three variables. If you start with $c = 3$, you probably want the top of the fraction to be a multiple of 3, such as 12. Let $b = 7$ so that the top is 12. Now, solve the equation to find the value of $a$, which is $a = \frac{7+5}{3} = \frac{12}{3} = 4$.

Second, find a numerical answer. In this case, the question just asks for the value of $b$, which is 7.

Third, plug your numbers in to answer choices. You get:

(A) $\frac{ac}{5} = \frac{4 \times 3}{5} = \frac{12}{5}$

(B) $ac - 5 = (4)(3) - 5 = 12 - 5 = 7$

(C) $c(a - 5) = 3(4 - 5) = 3(-1) = -3$

(D) $(a - 5)(c - 5) = (4 - 5)(3 - 5) = (-1)(-2) = 2$

(E) $5ac = (5)(4)(3) = 60$

As you can see, the only answer that matches is (B).

## Quiz #1

1. Alice is twice as old as Brian and Cathy is 6 years younger than Brian. If Alice is $a$ years old, how old is Cathy in terms of $a$?

(A) $a + 6$
(B) $a - 6$
(C) $\frac{a-12}{2}$
(D) $\frac{a+12}{2}$
(E) $2a - 6$

2. All the widgets manufactured by Company X are stored in two warehouses, A and B. Warehouse A contains three times as many widgets as does warehouse B. If $\frac{1}{15}$ of the widgets in warehouse A and $\frac{1}{20}$ of the widgets in warehouse B are defective, what fraction of all the widgets are not defective?

(A) $\frac{1}{75}$
(B) $\frac{1}{16}$
(C) $\frac{15}{16}$
(D) $\frac{34}{35}$
(E) $\frac{74}{75}$

3. Of all the players in a professional baseball league, $\frac{1}{2}$ are foreign-born, including $\frac{1}{3}$ of the pitchers. If $\frac{3}{4}$ of the players are pitchers, what percentage of the players who are not pitchers are foreign-born?

(A) 100%
(B) 75%
(C) $66\frac{2}{3}$%
(D) 50%
(E) 25%

## Quiz #1 — Answers and Explanations

1. (C) is correct. First, make up a number for $a$, such as $a = 20$. Second, solve for a numerical answer. If Alice is 20, then Brian is 10. If Brian is 10, then Cathy is $10 - 6 = 4$. So the answer is 4. Third, plug $a = 20$ into the answers and find the one that equals 4. The only one that matches is (C).

2. (C) is correct. First, make up some numbers for the number of widgets in A and B. Suppose A contains 300 widgets and B contains 100 widgets. Second, find a numerical answer. Warehouse A contains $\frac{1}{15} \times 300 = 20$ defective widgets. Warehouse B contains $\frac{1}{20} \times 100 = 5$ defective widgets. The total number of widgets is $100 + 300 = 400$ and the number of defective widgets is $20 + 5 = 25$. So the number of non-defective widgets is $400 - 25 = 375$. The fraction of widgets that are not defective is $\frac{375}{400} = \frac{15}{16}$. Choose (C).

3. (A) is correct. First, plug in a number for the total number of players. The number should be compatible with $\frac{1}{2}$, $\frac{1}{3}$, and $\frac{3}{4}$, so try 240 players. Second, solve for a numerical answer. You may want to use a grid like the ones you learned about in Chapter 6. See the example below. If there are 240 players total, then there are $\frac{1}{2} \times 240 = 120$ foreign-born players and $240 - 120 = 120$ native-born players. There are $\frac{3}{4} \times 240 = 180$ pitchers and $240 - 180 = 60$ non-pitchers. Of the pitchers, $\frac{1}{3} \times 180 = 60$ are foreign-born, leaving $180 - 60 = 120$ native-born pitchers. Of the 120 foreign-born players, 60 are pitchers, leaving $120 - 60 = 60$ foreign-born non-pitchers. So of the 60 non-pitchers, 60 are foreign-born. That's 100%. Choose (A).

| | Foreign | Native | Total |
|---|---|---|---|
| Pitchers | 60 | 120 | 180 |
| Non-pitchers | 60 | 0 | 60 |
| Total | 120 | 120 | 240 |

## BACKSOLVING

Just as Plugging In does, Backsolving also turns algebra problems into arithmetic problems. You use Plugging In when the answer choices contain variables—visible or invisible. Use Backsolving when the answer choices are numbers. Essentially, you try the answer choices to see which one fits with the information in the question. Here's the step-by-step method:

1. Identify the specific number for which the question asks. In other words, what value do the answer choices represent?
2. Try answer (C). Work through the problem using that number. If it works you're done. Otherwise:
3. Try a different answer choice. If answer (C) was too big, try a smaller answer. If (C) was too small, try a bigger number. Repeat until you find the answer that fits the information in the question.

Try this example:

1. Oliver has twice as many marbles as Ted. Ted has twice as many marbles as Merrill. If the total number of marbles among the three is 28, how many marbles does Ted have?

   (A) 4
   (B) 8
   (C) 12
   (D) 14
   (E) 16

First, identify the specific number asked for in the question. The question is "how many marbles does Ted have?" so the answer choices represent Ted's marbles.

Second, use answer (C) and work through the question. If Ted has 12 marbles, then Oliver has twice that, or $2 \times 12 = 24$ marbles. Merrill has half as many marbles as Ted, or $\frac{12}{2} = 6$ marbles. The total is 12 + 24 + 6 = 42 marbles. The total is supposed to be 28, so eliminate (C).

Third, try a different answer. The total from (C), 42, was too big, so you should look for a smaller number. Try (B). If Ted has 8 marbles, then Oliver has $2 \times 8 = 16$ marbles. Merrill has $\frac{8}{2} = 4$ marbles. The total is 8 + 4 + 16 = 28. That matches the number in the question, so (B) is the correct answer.

The Backsolving technique works when the question asks for a single number such as "What is $x$?" You cannot use it when the question asks for multiple numbers such as "What is $x + y$?" That's because you won't know how much of the answer is $x$ and how much is $y$.

It may take a little practice to get the hang of Plugging In and Backsolving, but they are invaluable for simplifying algebra questions, particularly "story problems." Take the time to get familiar with these techniques.

## Quiz #2

1. Rob is twice as old as Jodie is now. In 10 years, Rob will be 20 years older than Jodie is at that time. How old is Jodie now?

   (A) 10
   (B) 20
   (C) 30
   (D) 40
   (E) 50

2. John and Mark each own a collection of baseball cards. The two collections combined contain 120 cards. If John were to trade 5 cards to Mark and receive 2 of Mark's cards in return, John would have 22 more cards than Mark does. How many cards does Mark possess before the proposed trade?

   (A) 85
   (B) 74
   (C) 52
   (D) 46
   (E) 35

## Quiz #2—Answers and Explanations

1. (B) is correct. Start with answer (C). If Jodie is 30 years old now, then Rob is $2 \times 30 = 60$ years old. In 10 years, Rob will be $60 + 10 = 70$ and Jodie will be $30 + 10 = 40$ years old. However, Rob will be $70 - 40 = 30$ years older than Jodie, not the stated 20 years. Try a different answer. In (B), if Jodie is 20 years old, then Rob is $2 \times 20 = 40$. In 10 years, Rob will be $40 + 10 = 50$ and Jodie will be $20 + 10 = 30$. Rob will be $50 - 30 = 20$ years older, so (B) is correct.

2. (D) is correct. First, figure out what the answer represents. In this case, it is the number of cards Mark owns at the start. Second, use (C) and work through the problem. If Mark owns 52 cards, then John owns $120 - 52 = 68$ cards. John then trades 5 cards to Mark and gets 2 back. That's a net of 3 cards to Mark. So John now has $68 - 3 = 65$ and Mark has $52 + 3 = 55$. That's a difference of 10 cards, not 22 cards. Eliminate (C). Third, try another answer. In (B), Mark starts with 74 cards, so John has $120 - 74 = 46$ cards. After the trade, John has $46 - 3 = 43$ cards and Mark has $74 + 3 = 77$ cards. John has *fewer* cards than Mark, so that's definitely the wrong direction. Eliminate (B) and (A) and try (D). Mark has 46 cards, so John has $120 - 46 = 74$ cards. After the trade, John has $74 - 3 = 71$ cards and Mark has $46 + 3 = 49$ cards. That's a difference of $71 - 49 = 22$ cards, which matches the information in the question. Choose (D).

## ***MUST BE* PROBLEMS**

Some questions will ask "what must be true?" or some variation on that. They supposedly test your knowledge of the properties of various types of numbers (e.g., odd, even, positive, negative, etc.). However, you can easily solve these with the Plugging In technique. The difference is that you need to Plug In more than once.

Try numbers from both sides of whichever issue the question is testing. For example, if the question is testing odd vs. even, plug in both an odd number and an even number. After you plug in an odd number, eliminate any answers that aren't true. Then plug in an even number and eliminate any answers that aren't true. In most cases, you'll be left with one answer at this point. If not, continue plugging in different numbers until you're left with one answer.

If you need to plug in more numbers, try one and zero. Earlier, you avoided those numbers because they did weird things. Now, however, you want the answers to do weird things so that you can eliminate them.

Look at this next example:

1. If the product $xy$ is negative, which of the following must be true?

   (A) $x > y$
   (B) $x < 0$
   (C) $y < 0$
   (D) $\frac{x}{y} < 0$
   (E) $x + y < 0$

The question is testing positive vs. negative, so you'll want to try both kinds of numbers. You are constrained by the fact that $xy$ is negative. The numbers you plug in must fit that condition. You might start with $x = 2$ and $y = -3$. (A) is false, so you can eliminate it. (B) is false, so eliminate it. (C) is true in this case, so keep it. (D) is true in this case, so keep it. (E) is true in this case, so keep it.

Next, try a different set of numbers. Try reversing which variables are positive and negative. Let $x = -2$ and $y = 3$. You only need to check (C), (D), and (E) because you've already eliminated the others. (C) is false, so eliminate it. (D) is true in this case, so keep it. (E) is false, so eliminate it. The correct answer must be (D).

## Quiz #3

1. If $x$ is an integer, which of the following must be odd?

   (A) $3x$
   (B) $2x$
   (C) $3x + 1$
   (D) $4x$
   (E) $4x + 1$

2. If $p$, $q$, and $r$ are nonzero numbers and $p = q - r$, which of the following must equal 0?

   (A) $p - r$
   (B) $\frac{p+r-1}{q}$
   (C) $\frac{p+q}{r}-1$
   (D) $p - (q + r)$
   (E) $\frac{p+r}{q}-1$

## Quiz #3—Answers and Explanations

1. (E) is correct. The issue is odd vs. even, so plug in numbers of both types. Let $x = 2$, an even number. (A) is 6, an even number, so eliminate it. (B) is 4, an even number, so eliminate it. (C) is 7, an odd number, so keep it. (D) is 8, an even number, so eliminate it. (E) is 9, an odd number, so keep it. Now try an odd number, such as $x = 3$. (C) is 10, an even number, so eliminate it. (E) is 13, an odd number, so (E) is the correct answer.

2. (E) is correct. There's not really an odd/even or positive/negative issue, so you'll just need to plug in different sets of numbers until you're left with one answer. Try $p = 3$, $q = 6$, and $r = 3$. Answer (A) is $3 - 3 = 0$. Keep it. Answer (B) is $\frac{3+6-1}{3}=\frac{8}{3}$. Eliminate it. (C) is $\frac{3+6}{3}-1=3-1=2$. Eliminate it. (D) is $3 - (6 + 3) = -6$. Eliminate it. (E) is $\frac{3+3}{6}-1=1-1=0$. Keep it. Try a different set of numbers, such as $p = 1$, $q = 3$, and $r = 2$. Now (A) is $1-2=-1$. Eliminate it. (E) is $\frac{1+2}{3}-1=1-1=0$. The correct answer must be (E).

## PRACTICE SET

1. If Joe Bob was 25 years old 5 years ago, how old was he $x$ years ago?

   (A) $x - 30$
   (B) $x - 20$
   (C) $30 - x$
   (D) $20 - x$
   (E) $20 + x$

2. Increasing the original length of a racetrack by 15% and then increasing the new length by 10% is equivalent to increasing the original length by

   (A) 30.0%
   (B) 27.5%
   (C) 26.5%
   (D) 25.0%
   (E) 12.5%

3. At a certain bakery, $\frac{1}{4}$ of the cookies sold in one week were chocolate chip and $\frac{1}{5}$ of the remaining cookies sold were oatmeal raisin. If $x$ of the cookies sold were oatmeal raisin, how many were chocolate chip?

   (A) $\frac{1}{20}x$
   (B) $\frac{9}{20}x$
   (C) $\frac{4}{5}x$
   (D) $\frac{5}{4}x$
   (E) $\frac{5}{3}x$

4. If 50% of the money in a certain portfolio was invested in stocks, 20% in bonds, 15% real estate, and the remaining $37,500 in a money market fund, what was the total amount invested in the portfolio?

(A) $100,000
(B) $125,000
(C) $175,000
(D) $250,000
(E) $375,000

5. Of all the pies baked in a certain bakery, $\frac{1}{2}$ are apple pies, $\frac{1}{7}$ are cherry pies, $\frac{1}{4}$ are pecan pies, and the rest are coconut cream pies. If the combined number of pecan pies and coconut cream pies is 40, how many pies total did the bakery bake?

(A) 56
(B) 84
(C) 91
(D) 105
(E) 112

6. Bill buys two types of soda. He buys $m$ bottles of Brand A at $0.50 each. He buys $n$ bottles of Brand B at $0.60 each. What is Bill's average cost in cents for a bottle of soda, in terms of $m$ and $n$ ?

(A) $\frac{0.5m+0.6n}{m+n}$

(B) $\frac{m+n}{110}$

(C) $\frac{1.10}{m+n}$

(D) $\frac{50m+60n}{m+n}$

(E) $\frac{50m+60n}{mn}$

7. If $a$, $b$, and $c$ are nonzero integers, which of the following must be an integer?

(A) $\frac{a+b}{c}+1$

(B) $abc - 1$

(C) $\frac{ab}{c}-1$

(D) $\frac{a}{b+c}+1$

(E) $\frac{1}{a}+\frac{1}{b}+\frac{1}{c}$

8. In a certain school, $\frac{3}{5}$ of the students are boys and the rest are girls. Of the boys, $\frac{1}{4}$ play soccer. If the number of girls who play soccer equal the number of boys who play soccer, what fraction of the girls play soccer?

(A) $\frac{1}{10}$

(B) $\frac{3}{20}$

(C) $\frac{1}{4}$

(D) $\frac{1}{3}$

(E) $\frac{3}{8}$

9. Jason has a handful of dimes and quarters. There are a total of 22 coins. If the total value of the coins is $3.25, how many dimes does Jason have?

(A) 7
(B) 8
(C) 11
(D) 12
(E) 15

10. If $a$ and $b$ are both nonzero integers, which of the following must be positive?

I. $a^2 + b^2$

II. $a^2 - b^2$

III. $(a - b)^2$

(A) I only
(B) II only
(C) III only
(D) I and II
(E) I, II, and III

## Practice Set—Answers and Explanations

1. (C) is correct. Let $x = 10$. If Joe Bob was 25 years old 5 years ago, he must be 30 years old now. So 10 years ago, he was 20 years old. Plug $x = 10$ into the answers and see which one matches. The answer must be (C).

2. (C) is correct. Plug In a number for the original length, say 100 yards. Increasing the length by 15% adds 15 yards for a new total of 100 + 15 = 115 yards. Increasing this new length by 10% adds $0.10 \times 115 = 11.5$ yards for a total length of 115 + 11.5 = 126.5 yards. That's an overall increase of 26.5 yards or $\frac{26.5}{100} = 26.5\%$. Choose (C).

3. (E) is correct. Suppose that the bakery sold a total of 20 cookies. Then there are $\frac{1}{4} \times 20 = 5$ chocolate chip cookies, leaving 20 – 5 = 15 cookies. Of these remaining cookies, $\frac{1}{5} \times 15 = 3$ are oatmeal raisin, so $x = 3$. The numerical answer for the number of chocolate chip cookies is 5. Plug $x = 3$ into the answers and see which one matches. The answer is (E).

4. (D) is correct. Start with (C). If there is $175,000 in the portfolio, there is 0.5 × $175,000 = $87,000 in stocks, 0.2 × $175,000 = $35,000 in bonds, 0.15 × $175,000 = $26,250 in real estate, and $175,000 – $87,500 – $35,000 – $26,250 = $26,250 left for the money market fund. That doesn't match the $37,500 stated in the question, so eliminate (C). You need a bigger number, so try (D). If the total amount is $250,000, then there is 0.5 × $250,000 = $125,000 in stocks, 0.2 × $250,000 = $50,000 in bonds, 0.15 × $250,000 = $37,500 in real estate, and $250,000 – $125,000 – $50,000 – $37,500 = $37,500 left for the money market fund. That matches the information in the question, so choose (D).

5. (E) is correct. Normally you'd start by checking (C). However, (C) and (D) are both odd. Because apple pies are $\frac{1}{2}$ of the total, you can't have an odd total. Eliminate (C) and (D). Now check the middle answer of what's left, (B). If there are 84 pies total, there are $\frac{1}{2}\times 84 = 42$ apple pies, $\frac{1}{7}\times 84 = 12$ cherry pies, and $\frac{1}{4}\times 84 = 21$ pecan pies. That leaves 84 – 42 – 12 – 21 = 9 coconut cream pies. The total of pecan and coconut cream pies is 21 + 9 = 30 pies. That's too few. Eliminate (B) and try a bigger answer. In (E), there are 112 pies total. So there are $\frac{1}{2}\times 112 = 56$ apple pies, $\frac{1}{7}\times 112 = 16$ cherry pies, and $\frac{1}{4}\times 112 = 28$ pecan pies. That leaves 112 – 56 – 16 – 28 = 12 coconut cream pies. The total of pecan and coconut cream pies is 28 + 12 = 40. That matches the information in the question, so choose (E).
6. (D) is correct. Plug In numbers for $m$ and $n$. Let $m$ = 10 bottles at 50 cents each for a cost of 10 × 50 = 500 cents. Let $n$ = 15 bottles at 60 cents each for a cost of 15 × 60 = 900 cents. Bill's total cost is 500 + 900 = 1400 cents for 10 + 15 = 25 bottles of soda. So his average cost is $\frac{1400}{25} = 56$ cents per bottle. Plug $m$ = 10 and $n$ = 15 into the answers to see which one equals 56. Remember you need an answer in cents, not in dollars. (A) is 0.56, that's the trap answer in dollars. (B) is $\frac{25}{110}$. (C) is $\frac{1.1}{25}$. (D) is 56. (E) is $\frac{1400}{110}$, which is about 13. Choose (D).
7. (B) is correct. Plug in some numbers, such as $a$ = 2, $b$ = 3, and $c$ = 4. (A) is $\frac{2+3}{4}+1=\frac{5}{4}+1=\frac{9}{4}$. That's not an integer, so eliminate it. (B) is (2)(3)(4) –1 = 24 – 1 = 23. That is an integer, so keep it. (C) is $\frac{2\times 3}{4}+1=\frac{6}{4}+1=\frac{10}{4}=\frac{5}{2}$. That's not an integer, so eliminate it. (D) is $\frac{2}{3+4}+1=\frac{2}{7}+1=\frac{9}{7}$. That's not an integer, so eliminate it. (E) is $\frac{1}{2}+\frac{1}{3}+\frac{1}{4}=\frac{13}{12}$. That's not an integer, so eliminate it. Choose (B).

8. (E) is correct. Make up a number for the total number of students, say 200. There are $\frac{3}{5} \times 200 = 120$ boys and $200 - 120 = 80$ girls. There are $\frac{1}{4} \times 120 = 30$ soccer players among the boys, so there are also 30 soccer players among the girls. The fraction of girls that play soccer is $\frac{30}{80} = \frac{3}{8}$. Choose (E).

9. (E) is correct. Start Backsolving with (C). He has 11 dimes worth $11 \times 0.10 =$ \$1.10. He has $22 - 11 = 11$ quarters worth $11 \times 0.25 =$ \$2.75. That's a total value of $1.10 + 2.75 =$ \$3.85. That doesn't match the \$3.25 from the question. Eliminate (C). You want less money, so you need fewer quarters and more dimes. Try (D). Jason has 12 dimes worth $12 \times 0.10 =$ \$1.20. He has $22 - 12 = 10$ quarters worth $10 \times 0.25 =$ \$2.50. That's a total value of $1.20 + 2.50 =$ \$3.70, which is still too big. Try (E). Jason has 15 dimes worth $15 \times 0.10 =$ \$1.50. He has $22 - 15 = 7$ quarters worth $7 \times 0.25 =$ \$1.75. That's a total value of $1.50 + 1.75 =$ \$3.25, which matches the information in the question. Choose (E).

10. (A) is correct. Plug In numbers for $a$ and $b$, such as $a = 2$ and $b = 3$. I is $4 + 9 = 13$. That's positive, so keep it. II is $4 - 9 = -5$. That's negative, so you can eliminate it. That means you eliminate (B), (D), and (E). III is $(-1)^2 = 1$. That's positive, so keep it. Try strange numbers such as $a = 2$ and $b = 2$. I is $4 + 4 = 8$. That's positive, so keep it. III is $(0)^2 = 0$. That's not positive, so eliminate it and (C). The answer must be (A).

# 9

# Geometry

Although you probably studied geometry in high school, the geometry tested on the GMAT is very different. There are no proofs to memorize and regurgitate. However, there are some rules that you need to know and, as always, the GMAT writers will try to trick and confuse you.

The diagrams on Problem Solving geometry questions are drawn roughly to scale, unless they say otherwise. This means that an angle that looks significantly bigger than another angle probably is. However, you should know that the computer screen may occasionally distort the diagrams a bit. Don't assume that an angle that looks like a 90 degree angle really is one. It might be 89 or 91 degrees. Also, straight lines sometimes look jagged on the screen.

The diagram on a Data Sufficiency question is a different story. *Never* assume that a Data Sufficiency diagram is drawn to scale.

## ANGLES AND LINES

Angles come in three flavors. **Acute angles** are angles of less than 90 degrees. **Obtuse angles** are angles of more than 90 degrees. **Right angles** are angles of exactly 90 degrees. A straight **line** makes an angle of 180 degrees.

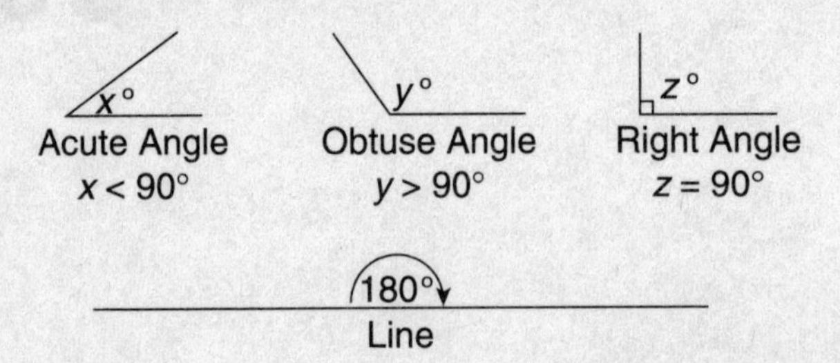


When two lines intersect, they form two kinds of angles. **Supplementary angles** are angles that combine to form a line. Therefore, they must add up to 180 degrees. **Vertical angles** are the angles across from each other. They are equal. See the diagram below:

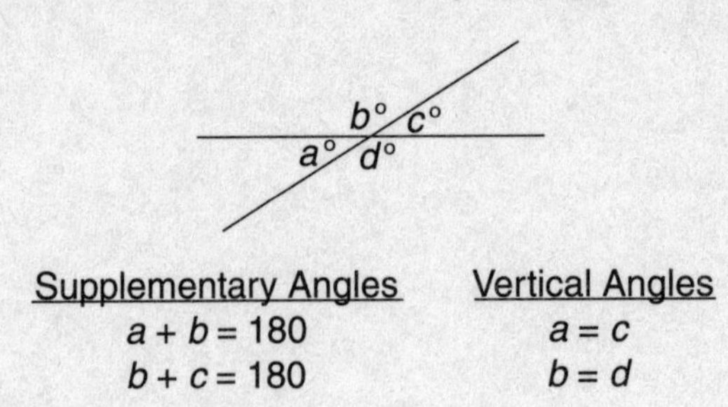

**Parallel lines** are lines that never intersect. When two parallel lines are intersected by a third line, eight angles are created. However, these angles only come in two sizes. All of the big angles are equal and all of the small angles are equal. See the diagram below:

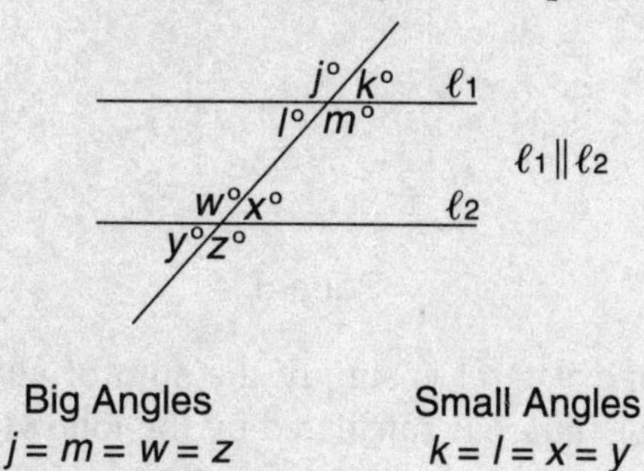


Big Angles
$j = m = w = z$

Small Angles
$k = l = x = y$

## QUADRILATERALS

The term **quadrilateral** refers to any four-sided figure. The only rule that quadrilaterals follow is that the four angles inside one must add up to 360 degrees. There are three specific types of quadrilaterals that do follow other rules: the parallelogram, the rectangle, and the square.

**Parallelograms** are composed of two sets of parallel lines. Because they involve parallel lines, parallelograms follow the rule about two sizes of angles: The big angles are equal and the small angles are equal. Additionally, each side is equal in length to the side opposite it.

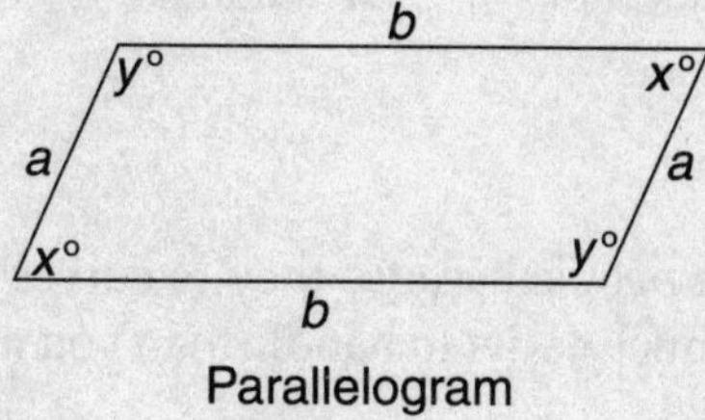


Parallelogram

**Rectangles** are quadrilaterals with four right angles. This also means that their opposite sides are equal. Every rectangle is also a parallelogram.

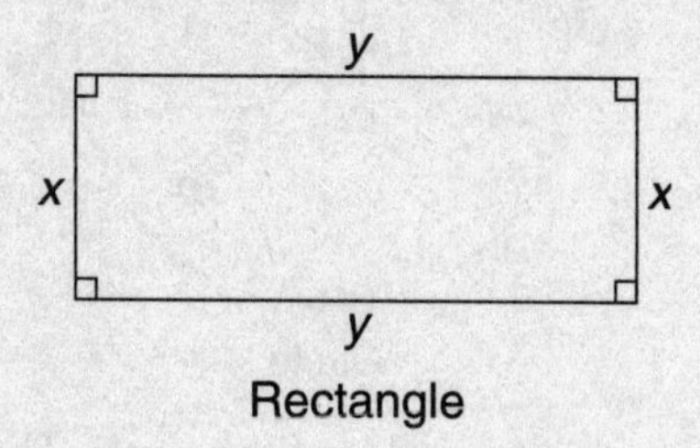


Rectangle

**Squares** are quadrilaterals that have four right angles and four equal sides. Every square is also a rectangle and a parallelogram.

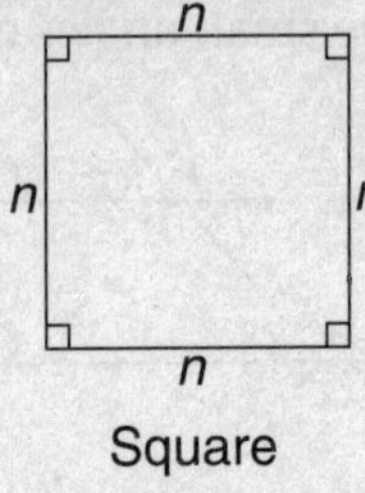


Square

The **perimeter** of a quadrilateral is simply the sum of all four sides. The **area** of a parallelogram, rectangle, or square is calculated by the following formula: *Area* = *Base* × *Height*. For rectangles, the area is sometimes expressed as *Area* = *Length* × *Width*. For squares, the area is sometimes expressed as *Area* = $Side^2$. However, these are just different ways of saying the same thing. The one important thing to remember is that the base and height must be **perpendicular** (forming a right angle). This is most important with parallelograms, because the sides are not necessarily perpendicular. In that case, the height is a line from the top to the base that forms a right angle to the base. See the diagram below:

Area = Base × Height

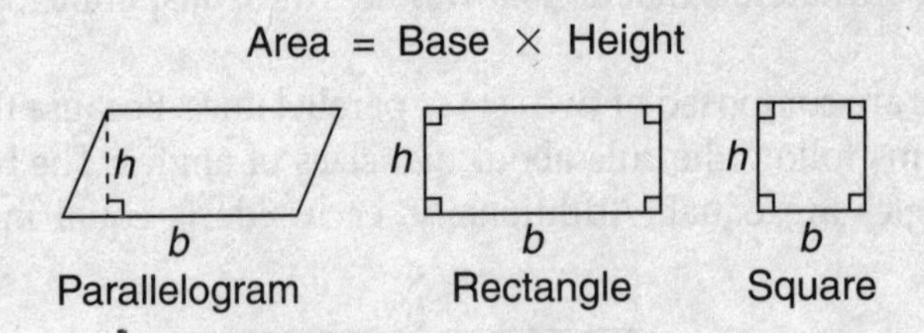


Parallelogram Rectangle Square

## CIRCLES

**Circles** sometimes intimidate people because they're curves that you can't measure with a ruler. However, circles are much easier to handle than you might think. Take a look at the parts of a circle:

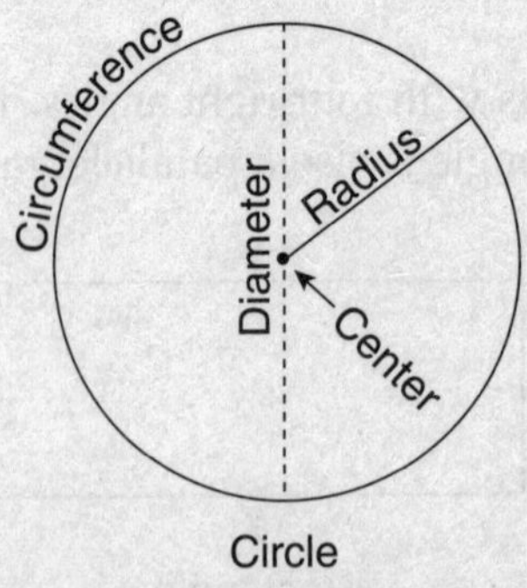


Circle

The **center** is the point at, well, the center of the circle. The **circumference** of the circle is the outside of the circle. It's very much like the perimeter of a quadrilateral or triangle. Each point on the circumference is the same distance from the center of the circle. The distance from the center of the circle to the circumference is the **radius**. It's a very important part of the circle because it plays a role in measuring all the other parts of the circle. A **diameter** is a line that runs from the edge of the circle, through the center, to the edge on the other side. The diameter is twice the length of the radius and it is the longest possible distance within a circle. All circles measure 360 degrees.

The formula for the **area** of a circle is *Area* = $\pi r^2$, where *r* is the radius of the circle. $\pi$ is the symbol for pi, a constant. You don't need to know the exact value of pi, just that it's a little more than 3. For example, if the radius of a circle is 3, the area is roughly $\pi(3^2) = 9\pi$. The formula for the **circumference** of circle is *Circumfrence* = $2\pi r$, where *r* is the radius of the circle. For example, if the radius of a circle is 3, the circumference is $(2)(\pi)(3) = 6\pi$. As you can see, finding the radius of the circle is usually the first step in solving a circle problem.

Sometimes the question will ask about a piece of a circle, like a slice of a pie. The rule for these situations is that all the characteristics of the slice are proportional to the size of the slice. For example, if the slice is $\frac{1}{4}$ of the circle (like *AOB* in the diagram above), then the area of the slice is $\frac{1}{4}$ of the area of the whole circle. The **arc** of the slice (the piece of the circumference) is $\frac{1}{4}$ of the circumference of the whole circle. The angle of the slice is $\frac{1}{4}$ of the 360 degrees in a circle, or 90 degrees. If it's a pie, the slice has $\frac{1}{4}$ of the calories of the whole pie. A **chord** is the straight line between the two endpoints of an arc.

## Quiz #1

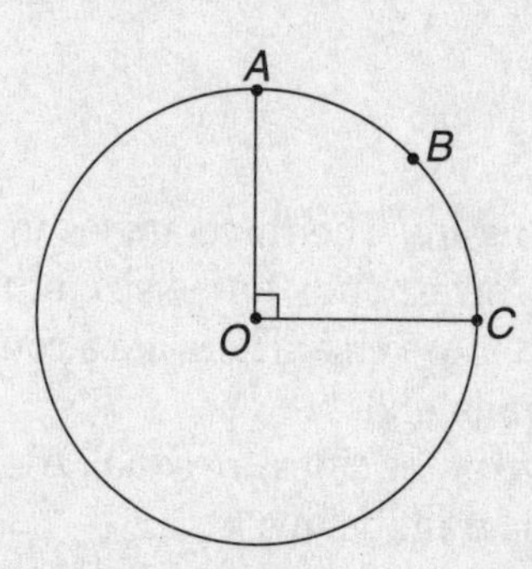


1. If the circle above has center *O* and an area of $36\pi$, what is the perimeter of sector *ABCO* ?

   (A) $6\pi$
   (B) $9\pi$
   (C) $6 + 3\pi$
   (D) $9 + 3\pi$
   (E) $12 + 3\pi$

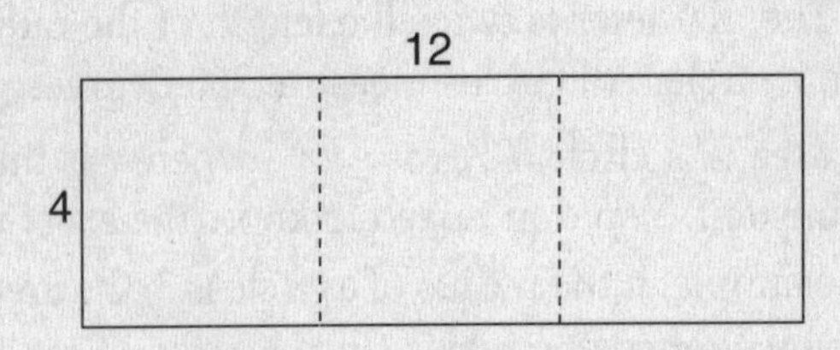


2. If a rectangle measuring 4 inches by 12 inches is cut into 3 equal rectangles as shown above, what is the perimeter, in inches, of each of the three new rectangles?

(A) $\frac{16}{3}$
(B) $\frac{32}{3}$
(C) 16
(D) 24
(E) 32

## Quiz #1 — Answers and Explanations

1. (E) is correct. If the area of the circle is $36\pi$, then the radius is 6. That means AO = 6 and CO = 6. The circumference of the circle is $2\times\pi\times6=12\pi$. Because the slice is 90 out of 360 degrees, it's a $\frac{90}{360}=\frac{1}{4}$ slice of the circle. So arc ABC is $\frac{1}{4}$ of the circumference or $\frac{1}{4}\times12\pi=3\pi$. So the perimeter of the slice is $6+6+3\pi=12+3\pi$. Choose (E).
2. (C) is correct. Each of the new pieces will be a square measuring 4 inches by 4 inches. So the perimeter will be 4 + 4 + 4 + 4 = 16 inches. The answer is (C).

# TRIANGLES

A **triangle** is any three-sided figure. The three angles in a triangle must add up to 180 degrees. The size of the sides corresponds to the size of the angles. The bigger sides are opposite the bigger angles, and vice versa. There are a couple of special triangles that you should know: isosceles and equilateral.

In an **isosceles triangle,** two of the sides are equal. Also, the two angles opposite those sides are also equal. See the diagram below:

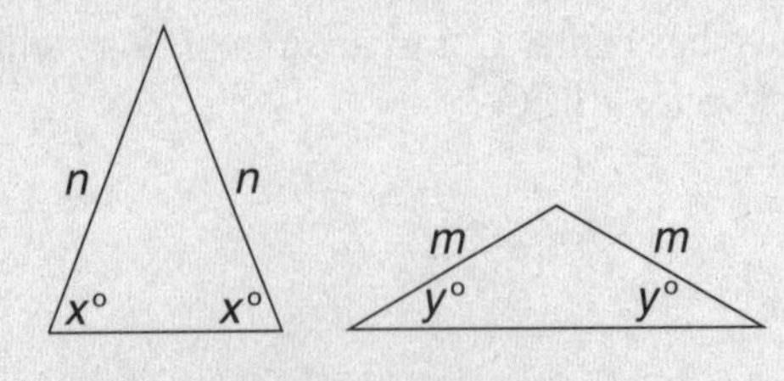


Isosceles Triangles

In an **equilateral triangle**, all three sides are equal. Also, all three angles are equal. Because there are 180 degrees in a triangle, each angle in an equilateral triangle measures $\frac{180}{3} = 60$ degrees. See the diagram below:

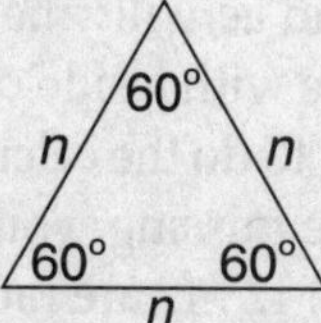


Equilateral Triangle

The **perimeter** of a triangle is simply the sum of the three sides. The **area** of a triangle is calculated by this formula: $Area = \frac{1}{2} \times Base \times Height$. As with quadrilaterals, the base must be perpendicular to the height. If two sides form a right angle, you can use those as the base and height. Otherwise, you'll have to draw in a line to serve as the height. See the diagram below:

$$\text{Area} = \frac{1}{2} \times \text{Base} \times \text{Height}$$

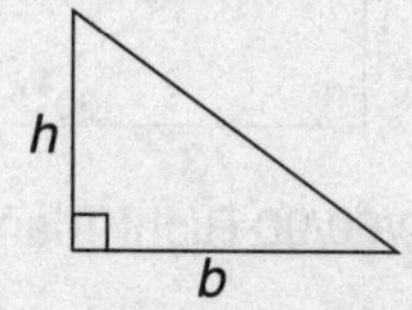


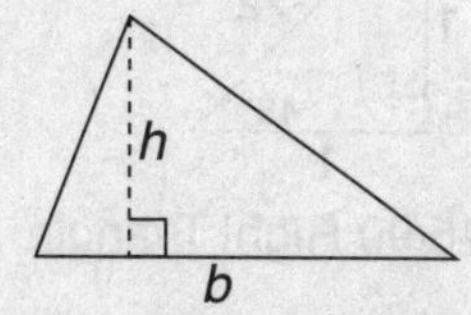


If one of the angles in a triangle is a right angle, that triangle is called a **right triangle**. The lengths of the sides of right triangles are related by a formula called the Pythagorean Theorem. If $a$, $b$, and $c$ are the sides of the triangle, then $a^2 + b^2 = c^2$. One critical element is that $c$ must be the **hypotenuse** (the longest side, opposite the right angle). Using this formula, you use the lengths of any two sides to calculate the length of the third side. Just plug the two known numbers into the appropriate places in the formula and solve for the remaining variable. Remember that the Pythagorean Theorem is only for *right* triangles, not all triangles.

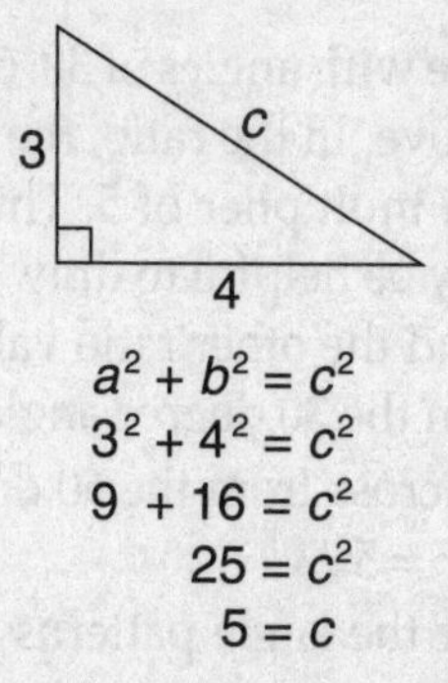


$$a^2 + b^2 = c^2$$
$$3^2 + 4^2 = c^2$$
$$9 + 16 = c^2$$
$$25 = c^2$$
$$5 = c$$

For example, in the figure above, the two known sides are 3 and 4. They go in the $a$ and $b$ spots in the formula because the other side, $c$, is the hypotenuse. Filled in, the formula now reads $3^2 + 4^2 = c^2$. When you solve the equation, you get $c = 5$.

The GMAT writers have a couple of favorite versions of the Pythagorean Theorem. They are the 3 : 4 : 5 right triangle and the 5 : 12 : 13 right triangle. These are nothing more than some numbers that happen to work nicely with the Pythagorean Theorem to result in integers. They're just shortcuts to avoid actually making the calculation with the formula. For example, if you see a right triangle with a side of 3 and a hypotenuse of 5, you can fill in 4 for the other side without needing to do the calculation. If you see a right triangle with sides of 5 and 12 with the hypotenuse missing, you can fill in 13 for the hypotenuse. It's important to realize that 3 : 4 : 5 and 5 : 12 : 13 are ratios. So a 30 : 40 : 50 triangle works just as well. If you see a right triangle with sides of 10 and 24 with the hypotenuse missing, you can fill in 26 because it's just a 5 : 12 : 13 right triangle with everything doubled.

There are two other cases of the Pythagorean Theorem that the GMAT frequently tests. These are the 45/45/90 right triangle and the 30/60/90 right triangle. The numbers refer to the angles in the triangle. The sides in a 45/45/90 right triangle must fit the ratio $1 : 1 : \sqrt{2}$. This is still the Pythagorean Theorem. You can plug those numbers into the formula to check. The sides in a 30/60/90 right triangle must fit the ratio $1 : \sqrt{3} : 2$. The diagram below illustrates these relationships:

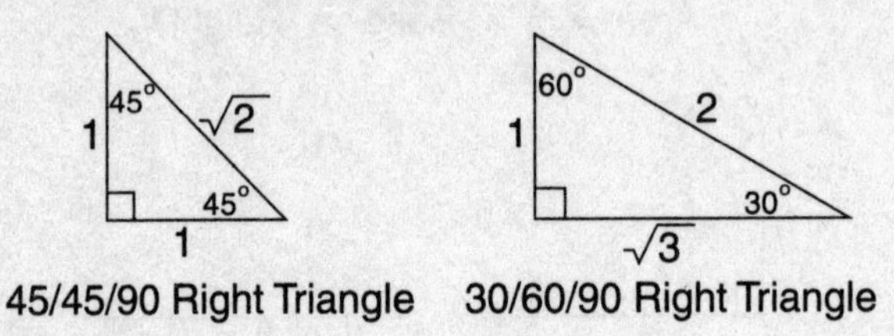


So why would you ever want to know this if you have the regular $a^2 + b^2 = c^2$ formula? Well, with the formula, you need numbers for two of the sides to be able to find the third side. With the ratios, however, you only need one of the sides to find the other two sides. Remember, a ratio and one actual value will tell you all of the other values.

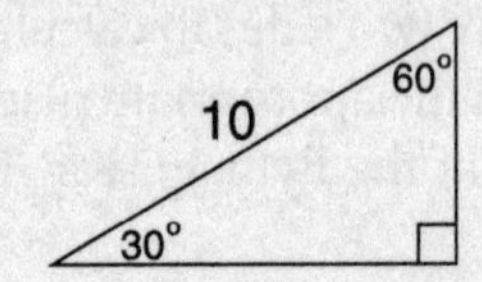


Suppose you see a right triangle with angles of 30, 60, and 90 degrees and a hypotenuse of 10, as shown in the diagram above. In the ratio, the hypotenuse has a value of 2. To get the actual value of 10, you need a multiplier of 5. This is just like the ratio methods you learned in Chapter 6. In fact, it may be helpful to draw ratio boxes to work these problems. You can then use the multiplier and the other ratio values to find the actual values for the missing sides. The side across from the 30 degree angle has a ratio value of 1, so the actual value must be $1 \times 5 = 5$. The side across from the 60 degree angle has a ratio value of $\sqrt{3}$, so the actual value must be $\sqrt{3} \times 5 = 5\sqrt{3}$.

Learn those ratios and look for the angle patterns that tell you to apply them.

## Quiz #2

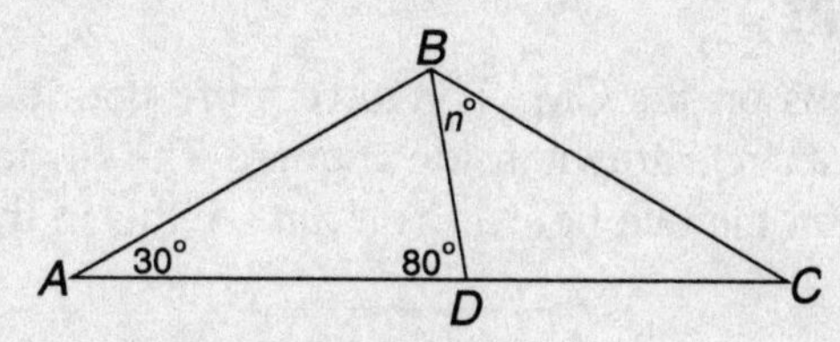


1. In the figure above, if $AB = BC$, then $n =$

   (A) 30°
   (B) 40°
   (C) 50°
   (D) 60°
   (E) 70°

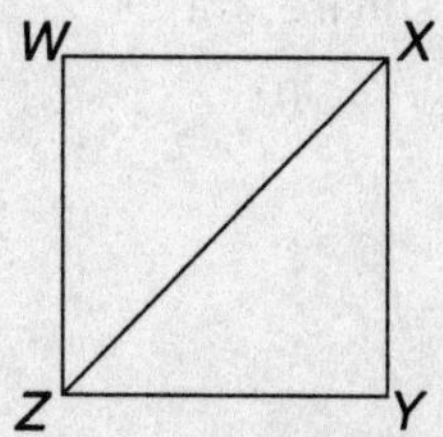


2. If square $WXYZ$ above has an area of 100, what is the length of diagonal $XZ$ ?

   (A) 10
   (B) $10\sqrt{2}$
   (C) $10\sqrt{3}$
   (D) $100\sqrt{2}$
   (E) $100\sqrt{3}$

## Quiz #2—Answers and Explanations

1. (C) is correct. If $AB = BC$, then the triangle is isosceles and angle $C$ equals angle $A$, measuring 30 degrees. Angles $ADB$ and $CDB$ form a straight line, so they must add up to 180 degrees. That means angle $CDB = 180 - 80 = 100$ degrees. You know that a triangle must add up to 180 degrees, so $n = 180 - 100 - 30 = 50$ degrees. The answer is (C).
2. (B) is correct. If the square has an area of 100, then each side of the square has a length of $\sqrt{100} = 10$. $WXZ$ is really a 45/45/90 right triangle, so you can apply the ratio. The ratio value of $WX$ is 1 and the actual value is 10. So the multiplier must be 10. The ratio value of WZ, the hypotenuse, is $\sqrt{2}$, so the actual value will be $\sqrt{2} \times 10 = 10\sqrt{2}$. Notice that you can get the same answer by just plugging $a = 10$ and $b = 10$ into the $a^2 + b^2 = c^2$ formula. However, you should also get familiar with the ratio method because there will be problems in which it may be the only thing you can use. Choose (B).

## OVERLAPPING FIGURES

Many geometry problems on the GMAT involve more than one geometric shape. For example, you might see a circle drawn inside a square. The key to solving these problems is to find the link between the two figures. You can see this in the example below:

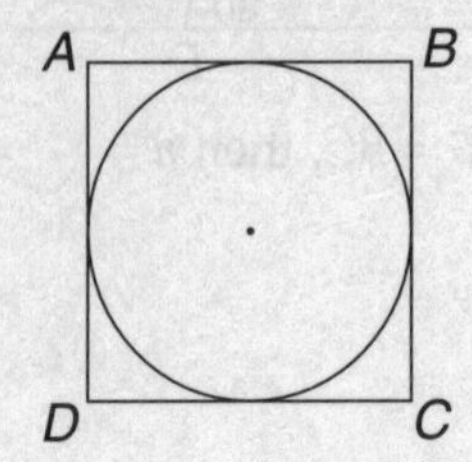


1. In the figure above, the circle with center $O$ has an area of $25\pi$. What is the area of square $ABCD$ ?

   (A) 5
   (B) 10
   (C) 25
   (D) 50
   (E) 100

If the circle has an area of $25\pi$, you can determine that the radius is 5, which makes the diameter 10. The link between the two shapes is that the diameter of the circle equals the side of the square. Try drawing a vertical diameter for the circle and seeing how it matches the vertical side of the square. So the side of the square is 10, which means that the area of the square is $10 \times 10 = 100$. The correct answer is (E).

If the problem involves the area of a shaded region, you can usually find that by subtracting the area of one figure from the area of another figure.

# SOLIDS

Solids are three-dimensional geometric shapes, including cubes, cylinders, spheres, and so on. You may be asked to find the volume of a solid or the surface area of a solid.

Finding the volume of a solid is very similar to finding the area of a shape. In fact, if you think of a solid as a flat shape times one more dimension, that's how you find the volume. Just multiply the area by the extra number. Suppose that you have a rectangle with measurements of 3 by 5. The area is $3 \times 5 = 15$. If you have a rectangular solid with measurements of 3 by 5 by 6, the volume is $3 \times 5 \times 6$. A cube is just a rectangular solid with the same measurement in all three dimensions.

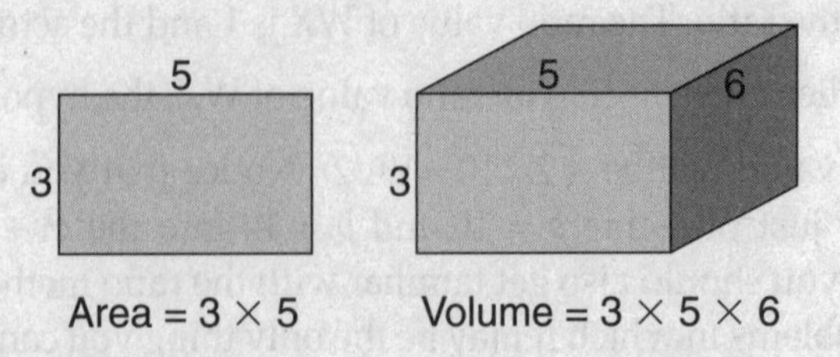

The volume of a cylinder follows a similar pattern. Start with the area of the circle. Then multiply by the height of the cylinder. Officially, the formula is $V = \pi \times r^2 \times h$, but that's just the area of the circle times *h*. For any other solid, such as a sphere, the question will provide the formula.

To calculate the surface area of a solid, break it up into several flat figures and find the area of each. For example, the 3 by 5 by 6 rectangular solid mentioned earlier has six faces: two are 3 by 5, two are 3 by 6, and two are 5 by 6. So the total surface area is $(3\times5) + (3\times5) + (3\times6) + (3\times6) + (5\times6) + (5\times6) = 15 + 15 + 18 + 18 + 30 + 30 = 126$.

A cylinder breaks up into two circles and one rectangle. One side of the rectangle is equal to the height of the cylinder. The other side is equal to the circumference of the circle. Imagine unrolling a paper towel on a cardboard tube.

## COORDINATE GEOMETRY

The coordinate grid is made up of an *x*-axis and a *y*-axis. When a coordinate is listed, it will be in the format of (*x*, *y*). The *x* number indicates the position on the *x*-axis and the *y* indicates the position on the *y*-axis. It's very easy to get your *x* and *y* confused. Be careful, because those reversed answers will usually be there to trap you. The diagram below demonstrates the coordinate grid and the position of a point:

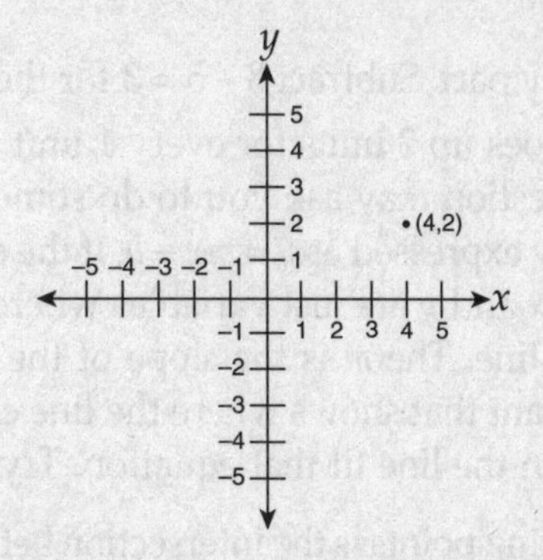


Some geometry questions will provide a line on a coordinate grid and ask you to determine its length. This is really just a triangle question in disguise. Draw a right triangle, making the line in question the hypotenuse. Make each of the other sides parallel to either the *x*-axis or the *y*-axis. Look at the example below:

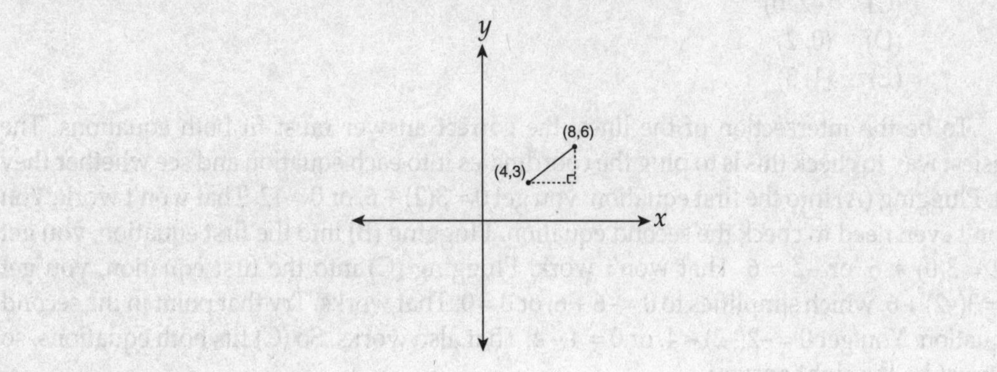

The horizontal side goes from $x = 4$ to $x = 8$, so its length is 4. The vertical side goes from $y = 3$ to $y = 6$, so its length is 3. You just use the Pythagorean Theorem to find the length of the third side. In this case, it's the hypotenuse of a 3 : 4 : 5 right triangle, so the length is 5.

Some questions may ask you to find the **slope** of a line. The easy definition of slope is "rise over run" or $\frac{\text{rise}}{\text{run}}$. Just find two points on the line and subtract one $y$ from the other and put the result on top of the fraction. Then subtract one $x$ from the other and put that on the bottom of the fraction. Make sure you start with the same point for both subtractions. Otherwise, the positive/negative sign of your answer will be wrong. Here's an example:

1. If the points (5, 7) and (8, 13) are on a line, what is the slope of that line?

   (A) $-2$

   (B) $-\frac{1}{2}$

   (C) $\frac{1}{2}$

   (D) 2

   (E) 8

Subtract $13 - 7 = 6$ for the $y$ part. Subtract $8 - 5 = 3$ for the $x$ part. So the slope is $\frac{6}{3} = 2$. The answer is (D). The line goes up 2 units for every 1 unit it goes over to the right.

On occasion, a GMAT question may ask you to do something with the equation of a line. A line equation is usually expressed as $y = mx + b$. If the equation is not in that format, rearrange it so that it is. The $x$ and $y$ are just variables where you plug in or solve for the coordinates of points on the line. The $m$ is the slope of the line that you just learned to calculate. The $b$ is just a constant that shows where the line crosses the $y$-axis. The $x$ and $y$ coordinates for every point on the line fit that equation. Try this example:

2. Which of the following points is the intersection between the lines $y = 3x + 6$ and $y = -2x - 4$?

   (A) (2, 0)

   (B) (0, –2)

   (C) (–2, 0)

   (D) (0, 2)

   (E) (1, 5)

To be the intersection of the lines, the correct answer must fit both equations. The easiest way to check this is to plug the coordinates into each equation and see whether they fit. Plugging (A) into the first equation, you get $0 = 3(2) + 6$, or $0 = 12$. That won't work. You don't even need to check the second equation. Plugging (B) into the first equation, you get $-2 = 3(0) + 6$, or $-2 = 6$. That won't work. Plugging (C) into the first equation, you get $0 = 3(-2) + 6$, which simplifies to $0 = -6 + 6$, or $0 = 0$. That works. Try that point in the second equation. You get $0 = -2(-2) - 4$, or $0 = 4 - 4$. That also works. So (C) fits both equations, so it must be the right answer.

## Quiz #3

1. A cylinder of wax is 20 inches tall with a radius of 1 inch. If this cylinder is melted and poured into a new cylindrical mold with a radius of 2 inches, what will be the height of the new cylinder in inches?

   (A) 5
   (B) 10
   (C) 15
   (D) 20
   (E) 40

2. A line that passes through the point (2, 5) has a slope of $-\frac{2}{3}$. Which of the following points are NOT on the line?

   (A) (11, –1)
   (B) (8, 1)
   (C) (4, 7)
   (D) (–1, 7)
   (E) (5, 3)

## Quiz #3—Answers and Explanations

1. (A) is correct. Find the volume of the original cylinder. It's $\pi \times r^2 \times h = \pi \times 1^2 \times 20 = 20\pi$. The volume doesn't change when the shape does, so plug that volume into the formula for the new cylinder: $v = \pi \times r^2 \times h$ becomes $20\pi = \pi \times 2^2 \times h$, so $h = 5$. Choose (A).
2. (C) is correct. You can start with (2, 5) and work your way down the line by adding –2 to the $y$ coordinate and 3 to the $x$ coordinate. This leads to points (5, 3), (8, 1), and (11, –1). You can also start at (2, 5) and work the other way by adding 2 to $y$ and –3 to $x$. This leads to (–1, 7). You can eliminate all those answers, leaving (C) as the only point not on the line. Choose (C).

Alternatively, you can use the given point, (2, 5), and the given slope, $m = -\frac{2}{3}$, to determine the line equation. Just plug those numbers into the format to get $5 = -\frac{2}{3}(2) + b$. You can solve that for $b = \frac{19}{3}$. Then you plug each point into the new line equation to see whether it fits. That way is more tedious, but will work.

# PRACTICE SET

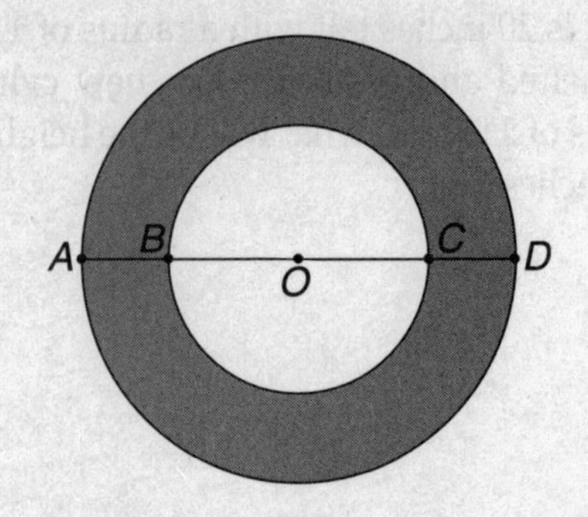


1. The figure above contains two circles with center $O$. If $OD = 10$ and the area of the shaded region is $36\pi$, what is the area of the smaller circle?

   (A) $6\pi$
   (B) $10\pi$
   (C) $36\pi$
   (D) $64\pi$
   (E) $100\pi$

2. If the length of a rectangle is decreased by 10 percent and its width is decreased by 20 percent, by what percent does its area decrease?

   (A) 30%
   (B) 28%
   (C) 25%
   (D) 23%
   (E) 15%

3. Freddy has a piece of string that's 20 inches long. He wants to use the string as the perimeter of a rectangle. What is the greatest area that rectangle could have?

   (A) 5
   (B) 10
   (C) 20
   (D) 25
   (E) 40

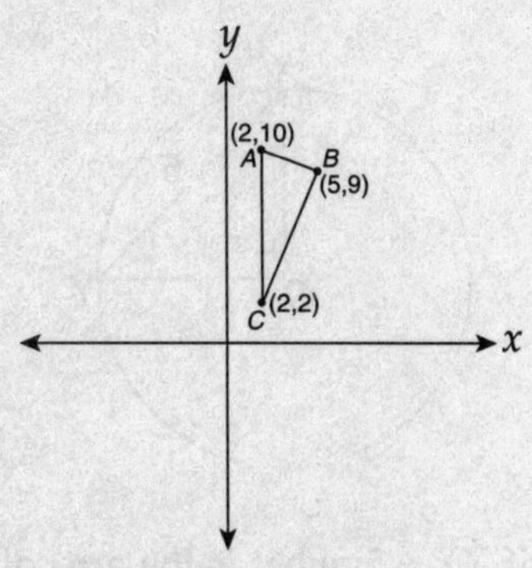


4. In the coordinate system above, what is the area of triangle *ABC* ?

   (A) 10
   (B) 12
   (C) 24
   (D) 25
   (E) 50

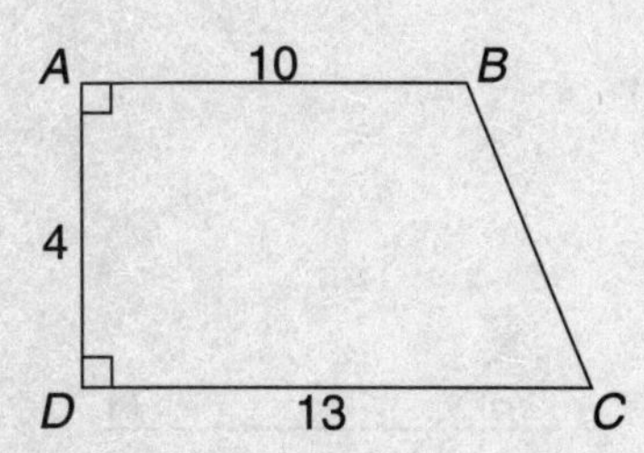


5. What is the area of quadrilateral *ABCD* above?

   (A) 52
   (B) 46
   (C) 40
   (D) 26
   (E) 20

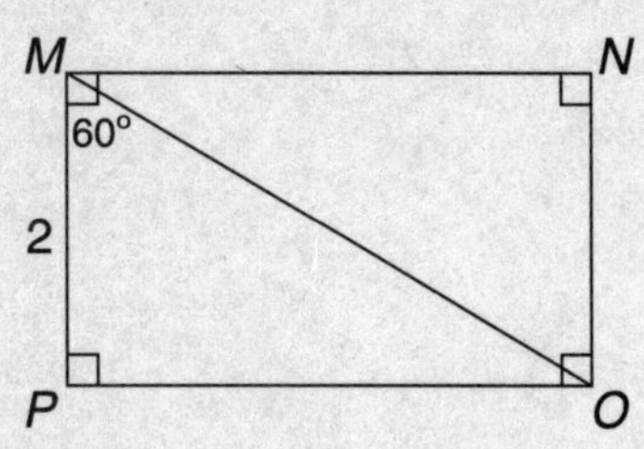


6. In rectangle *MNOP* above, if *MP* = 2, what is the length of *MO* ?

   (A) $2\sqrt{2}$
   (B) $2\sqrt{3}$
   (C) 4
   (D) 5
   (E) 6

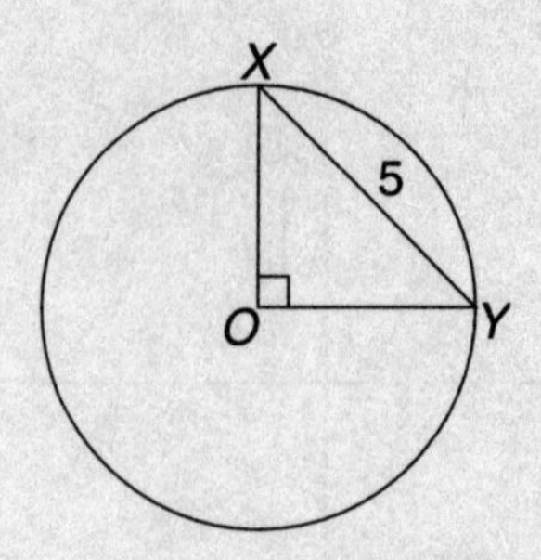


7. In the figure above, if $XY = 5$, what is the area of the circle with center $O$?

(A) $\frac{5\sqrt{2}}{2}\pi$
(B) $\frac{25}{4}\pi$
(C) $5\sqrt{2}\pi$
(D) $\frac{25}{2}\pi$
(E) $25\pi$

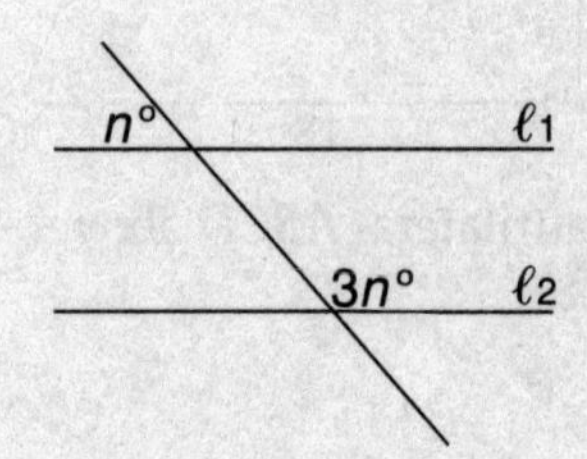


8. If $l_1$ is parallel to $l_2$ in the figure above, what is the value of $n$?

(A) 30
(B) 45
(C) 60
(D) 105
(E) 135

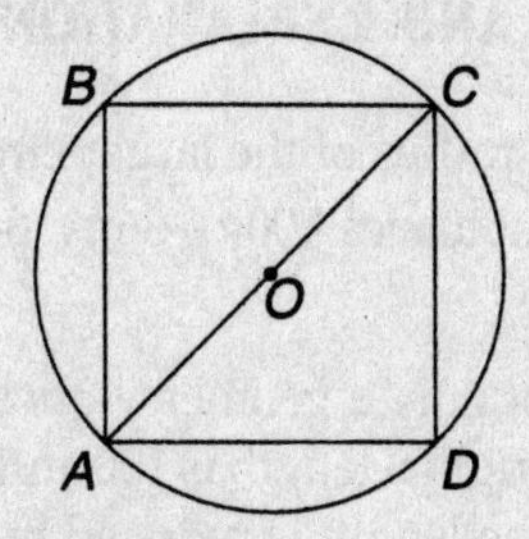


9. In the figure above, square $ABCD$ has an area of 25. What is the area of the circle with center $O$ ?

(A) $\frac{5\sqrt{2}}{2}\pi$

(B) $\frac{25}{4}\pi$

(C) $\frac{25}{2}\pi$

(D) $25\pi$

(E) $25\sqrt{2}\pi$

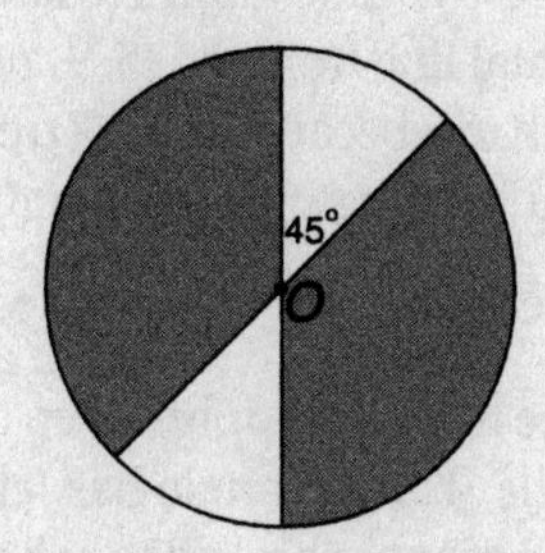


10. In the figure above, what fraction of the circle with center $O$ is shaded?

(A) $\frac{1}{8}$

(B) $\frac{1}{4}$

(C) $\frac{2}{3}$

(D) $\frac{3}{4}$

(E) $\frac{5}{6}$

## Practice Set — Answers and Explanations

1. (D) is correct. OD is the radius of the larger circle, so its area is $100\pi$. If the shaded region is $36\pi$, that leaves $100\pi - 36\pi = 64\pi$ for the area of the smaller circle. Choose (D).

2. (B) is correct. This is not only a geometry question, but also a Plugging In question. Suppose the original length is 100 and the original width is 50. The original area is 5,000. The length is reduced by 10% to 90 and the width is reduced by 20% to 40. The new area is 2,600. That's a decrease of $5{,}000 - 2{,}600 = 1{,}400$. To make that a percent change, divide by the original amount: $\frac{1{,}400}{5{,}000} = 28\%$. Choose (B).

3. (D) is correct. This is a good problem for Backsolving. Start with the largest answer and work your way down until you find one that works. (E) is too large. That rectangle would need length and width of 2 and 20, or 5 and 8. Those are too big for a perimeter of 20. (D) will work. The rectangle would have length and width of 5 and 5, which gives a perimeter of 20. Choose (D).

4. (B) is correct. Use the coordinates to determine the lengths of the necessary parts. $\text{Area} = \frac{1}{2} \times \text{Base} \times \text{Height}$. Since base needs to be perpendicular to height, you can't just use two sides of the triangle. Instead use the vertical side and draw a horizontal line from the (5, 9) point to the left. The base (the vertical line) has a length of 8 because it runs from $y = 2$ to $y = 10$. The height (the horizontal line you just drew) has a length of 3 because it goes from $x = 2$ to $x = 5$. So the area is $\frac{1}{2} \times 8 \times 3 = 12$. Choose (B).

5. (B) is correct. Draw a line from B straight down to split the quadrilateral into a rectangle and a triangle. The rectangle has length and width of 10 and 4, for a total area of 40. The triangle has sides of 3 and 4. They are perpendicular, so they can serve as base and height. That gives an area of 6. So the total area of the quadrilateral is $40 + 6 = 46$. Choose (B).

6. (C) is correct. Triangle *MOP* is a 30/60/90 right triangle, so use that ratio. The ratio value of *MP* (the side opposite the 30 degree angle) is 1 and the actual value is 2, so the multiplier must be 2. The ratio value of *MO*, the hypotenuse, is 2, so the actual value is $2 \times 2 = 4$. Choose (C).

7. (D) is correct. The key to overlapping figures is to find the piece that bridges the two figures. In this case, $OX$ and $OY$ are also radii of the circle, so they are equal, which makes $OXY$ a 45/45/90 right triangle. Use the ratios. The ratio value of XY, the hypotenuse, is $\sqrt{2}$ and the actual value is 5, so the multiplier must be $\frac{5}{\sqrt{2}}$. The ratio value for each of the other sides is 1, so the actual value is $1\times\frac{5}{\sqrt{2}}=\frac{5}{\sqrt{2}}$. Plug that into the formula for area of a circle to get $\pi\times r^2=\pi\left(\frac{5}{\sqrt{2}}\right)^2=\pi\left(\frac{25}{2}\right)=\frac{25}{2}\pi$. Choose (D).

8. (B) is correct. With a parallel line intersected by another line, only 2 sizes of angle are created. All of the big angles are equal to $3n$ and all of the small angles are equal to $n$. So $3n$ and $n$ form a straight line and $3n + n = 180$. That means that $4n = 180$ so $n = 45$. Choose (B).

9. (C) is correct. If the area of the square is 25, then each side of the square is 5. $ABC$ forms a 45/45/90 right triangle, and you know what that means . . . use the ratio. The ratio value of AB, one of the equal sides, is 1 and the actual value is 5, so the multiplier is 5. The ratio value of AC, the hypotenuse, is $\sqrt{2}$, so the actual value is $\sqrt{2}\times5=5\sqrt{2}$. The radius of the circle is half of AC, so the radius is $\frac{1}{2}\times5\sqrt{2}=\frac{5\sqrt{2}}{2}$. So the area of the circle is $\pi\times r^2=\pi\times\left(\frac{5\sqrt{2}}{2}\right)^2=\frac{25\times2}{4}\pi=\frac{25}{2}\pi$. Choose (C).

10. (D) is correct. The two unshaded regions are vertical angles, so they are equal, at 45 degrees each. Two 45 degree slices is like one 90 degree slice, which is $\frac{90}{360}=\frac{1}{4}$ of the circle, because it covers 90 of the 360 degrees in the circle. That leaves $1-\frac{1}{4}=\frac{3}{4}$ in the shaded region. Choose (D).

# 10

# More Data Sufficiency

In Chapter 3, you learned about the Data Sufficiency format and the "AD or BCE" method that you should use to tackle it. In this chapter, you will see some more sophisticated techniques that will help you with specific types of Data Sufficiency questions.

## SIMULTANEOUS EQUATIONS

In Chapter 7, you learned how to solve simultaneous equations that involve two variables. The general rule is: You need as many equations as you have variables. To solve for two variables, you need two equations. To solve for three variables, you need three equations.

For Data Sufficiency questions, you don't actually need to solve the equations; you just need to realize when you have enough information to do so. The key is recognizing equations and variables. In some cases, such as the following example, it's relatively easy.

1. If $x + y = 7$, what is $x$ ?

   (1) $x - y = 1$

   (2) $y + z = 11$

Statement (1) is sufficient because you now have two distinct equations. That's enough to solve for both variables and answer the question. Narrow it down to (A) or (D). Statement (2) is not sufficient, because it introduces a *third* variable and you only have *two* equations. You can't solve it, so (A) is the correct answer.

In other cases, identifying the variables and the equations becomes more difficult. Look at this next example:

2. Angela goes shopping. If she buys a total of 7 hats and belts, how many hats did she buy?

   (1) Angela buys one more hat than she does belts.

   (2) Angela buys a total of 11 belts and skirts.

In the question, there are two variables, the number of hats and the number of belts. There is also one equation, hats + belts = 7. In statement (1), there is one equation, hats – 1 = belts. In statement (2), there is one equation, belts + skirts = 11, and one more variable, the number of skirts. It's essentially the same question as #1, just disguised as a word problem. The correct answer is also (A). If you can spot the variables and equations in word problems, Data Sufficiency questions will become much easier.

### OVERLAPPING RANGES

In order to answer the question, the statement(s) must provide a single value for the number asked. If a statement narrows the possibilities down to a few numbers, but not just one, it is not sufficient to answer the question. However, if each statement narrows the possibilities to a small set and there is only one value in common, then those statements together are sufficient. Look at this example:

1. What is $x$ ?

   (1) $x$ is an odd integer between 0 and 10.

   (2) $x$ is a multiple of 5.

Statement (1) tells you that $x$ is 1, 3, 5, 7, or 9, however, that's not a single value, so you can't answer the question. Narrow the answers to (B), (C), or (E). Statement (2) tells you that $x$ is 5, 10, 15, 20, 25, or some other multiple of 5. However, that's not a single value. Eliminate (B). Both statements together are sufficient. The only value they share is 5, so $x$ must equal 5. Choose (C).

## QUIZ #1

1. What is the value of even integer $n$ ?

   (1) $\sqrt{n}$ is an integer.

   (2) $20 < n < 50$

   (A) Statement (1) ALONE is sufficient, but statement (2) alone is not sufficient to answer the question asked.
   (B) Statement (2) ALONE is sufficient, but statement (1) alone is not sufficient to answer the question asked.
   (C) BOTH statements (1) and (2) TOGETHER are sufficient to answer the question asked; but NEITHER statement ALONE is sufficient.
   (D) EACH statement ALONE is sufficient to answer the question asked.
   (E) Statements (1) and (2) TOGETHER are NOT sufficient to answer the question asked, and additional data specific to the problem are needed.

2. If a student's total cost for a semester's tuition, fees, and books was $11,250, how much was his cost for books that semester?

   (1) The cost for fees was 15 percent of the cost for tuition.

   (2) The combined cost for books and fees was 25 percent of the cost for tuition.

   (A) Statement (1) ALONE is sufficient, but statement (2) alone is not sufficient to answer the question asked.
   (B) Statement (2) ALONE is sufficient, but statement (1) alone is not sufficient to answer the question asked.
   (C) BOTH statements (1) and (2) TOGETHER are sufficient to answer the question asked; but NEITHER statement ALONE is sufficient.

(D) EACH statement ALONE is sufficient to answer the question asked.

(E) Statements (1) and (2) TOGETHER are NOT sufficient to answer the question asked, and additional data specific to the problem are needed.

## Quiz #1 — Answers and Explanations

1. (C) is correct. Look at statement (1). $\sqrt{n}$ is an integer such as 1, 2, 3, 4, 5, 6, 7, and so on. So $n$ must be a perfect square such as 1, 4, 9, 16, 25, 36, 49, and so on. That's not enough to answer the question, so narrow the choices to (B), (C), or (E). Look at statement (2). $n$ could be any even integer between 20 and 50, such as 22, 24, 26, etc. That's not enough to answer the question, so eliminate (B). Try (1) and (2) together. The only number on both lists is 36, so $n = 36$. You can answer the question, so choose (C).

2. (C) is correct. The question itself provides one equation and contains three variables. Look at statement (1). This gives you a second equation. However, you need three equations to solve for three variables. You can't answer the question, so narrow the choices to (B), (C), or (E). Look at statement (2). This provides another equation. However, you still have only two equations for three variables. You can't answer the question, so eliminate (B). Try (1) and (2) together. Now you have three distinct equations and you can solve for all three variables. You can answer the question, so choose (C).

## Pieces of the Puzzle

The answer to a Data Sufficiency problem revolves around which pieces of information are sufficient to answer the question posed. If you can determine which pieces of information are necessary, you can more easily recognize which statement(s) contain those pieces of information. Identify the value for which the question asks and determine which numbers you need to get there. It's like putting together a jigsaw puzzle.

Look at this next example:

1. If Steve owns red, blue, and green marbles in the ratio of $2:3:5$ respectively, how many blue marbles does he own?

   (1) Steve owns 20 red marbles.

   (2) Steve owns a total of 100 red, blue, and green marbles.

Before you even look at the statements, try to figure out what information you'll need to answer the question. You know that a ratio plus any of the actual values will allow you to find all the other actual values. So look for actual values as you check each statement.

Statement (1) contains an actual value, so it is sufficient. Narrow the choices to (A) or (D). Statement (2) has an actual value, so it is sufficient. The answer must be (D).

This approach works well with any Data Sufficiency problem involving a formula of some sort. For example, many geometry questions involve some sort of formula. Look at this example:

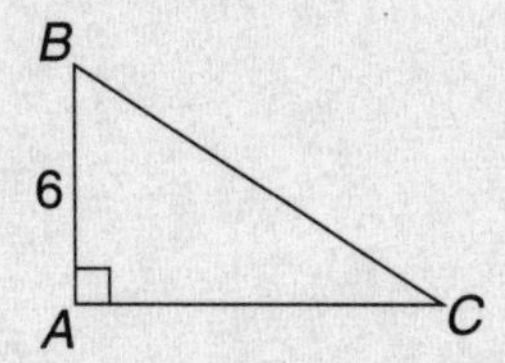


2. In triangle $ABC$ above, the length of $AB$ is 6. What is the area of triangle $ABC$ ?

   (1) The length of $BC$ is 10.

   (2) The length of $AC$ is 8.

To calculate the area of a triangle, you need to know the base and the height. The question provides the height, so the missing piece is the base, $AC$. A statement that gives you the length of $AC$ or allows you to calculate it is sufficient.

Statement (1) gives you the length of $BC$. Because ABC is a right triangle, you can use any two sides and the Pythagorean Theorem to find the length of the third side. With (1) you have two sides so you can find $AC$ and calculate the area. Narrow the choices to (A) or (D). Statement (2) tells you the length of $AC$. You can calculate the area, so choose (D).

This approach helps you to identify the missing piece so that you know what to look for, rather than working the problem from scratch for each statement.

## YES/NO QUESTIONS

Most of the Data Sufficiency questions you've seen so far ask for a specific value such as: "What is $x$ ?" However, you will probably see some questions that ask not for a numerical value but for a *yes* or *no* answer. For example, look at the question below:

1. Is $x$ an odd integer?

   (1) $x$ is divisible by 3.

   (2) $x$ is divisible by 2.

The question doesn't ask for the value of $x$. Instead it asks a yes/no question. A sufficient answer is either "Yes, $x$ is odd," or "No, $x$ is not odd." It's important to realize that "No" is an acceptable answer. The insufficient answer is "I can't tell from the information."

In the above question, look at statement (1). If $x$ is divisible by 3, it could be odd (such as 3, 9, or 15) or even (such as 6, 12, or 18). From statement (1) you cannot tell for certain whether $x$ is odd or not. You can't answer the question, so narrow the choices to (B), (C), or (E). From statement (2), you know that $x$ is even, so the answer to the question is "No, $x$ is not odd." You have answered the question, so choose (B). Remember, "No" is a sufficient answer.

## Quiz #2

1. Is it true that $x < y$ ?

   (1) $5x < 5y$

   (2) $xz < yz$

   (A) Statement (1) ALONE is sufficient, but statement (2) alone is not sufficient to answer the question asked.
   (B) Statement (2) ALONE is sufficient, but statement (1) alone is not sufficient to answer the question asked.
   (C) BOTH statements (1) and (2) TOGETHER are sufficient to answer the question asked; but NEITHER statement ALONE is sufficient.
   (D) EACH statement ALONE is sufficient to answer the question asked.
   (E) Statements (1) and (2) TOGETHER are NOT sufficient to answer the question asked, and additional data specific to the problem are needed.

2. If $p$ is a positive integer, is $p$ even?

   (1) $4p$ is even.

   (2) $p + 2$ is odd.

   (A) Statement (1) ALONE is sufficient, but statement (2) alone is not sufficient to answer the question asked.
   (B) Statement (2) ALONE is sufficient, but statement (1) alone is not sufficient to answer the question asked.
   (C) BOTH statements (1) and (2) TOGETHER are sufficient to answer the question asked; but NEITHER statement ALONE is sufficient.
   (D) EACH statement ALONE is sufficient to answer the question asked.
   (E) Statements (1) and (2) TOGETHER are NOT sufficient to answer the question asked, and additional data specific to the problem are needed.

## Quiz #2—Answers and Explanations

1. (A) is correct. Start with statement (1). If you divide both sides of the inequality by 5, you get $x < y$. So the answer to the question is "yes." Narrow the choices to (A) or (D). Look at statement (2). The problem here is that you don't know what kind of number $z$ is. If you plug in a positive number, such as $z = 4$, you get $x < y$ when you divide both sides by $z$. If you plug in a negative number, such as $z = -3$, then you get $x > y$ when you divide both sides by $z$. Remember, you have to flip the inequality sign when you multiply or divide by a negative number. You can't answer the question, so choose (A).

2. (B) is correct. Look at statement (1). $p$ could be either an odd number, such as $p = 3$, or an even number, such as $p = 2$. Both numbers fit the statement, so you can't answer the question. Narrow the choices to (B), (C), or (E). Look at statement (2). You can plug in odd numbers such as $p = 3$. However, you can't plug in even numbers because they won't fit the statement. So $p$ is odd and the answer is "no." You can answer the question, so choose (B).

# PRACTICE SET

1. What is the value of $n$ ?

   (1) $n^2 + 5n + 6 = 0$

   (2) $n^2 - n - 6 = 0$

   (A) Statement (1) ALONE is sufficient, but statement (2) alone is not sufficient to answer the question asked.
   (B) Statement (2) ALONE is sufficient, but statement (1) alone is not sufficient to answer the question asked.
   (C) BOTH statements (1) and (2) TOGETHER are sufficient to answer the question asked; but NEITHER statement ALONE is sufficient.
   (D) EACH statement ALONE is sufficient to answer the question asked.
   (E) Statements (1) and (2) TOGETHER are NOT sufficient to answer the question asked, and additional data specific to the problem are needed.

2. If $x$ and $y$ are integers, does $x = y$ ?

(1) $xy = y^2$

(2) $x^2 = y^2$

(A) Statement (1) ALONE is sufficient, but statement (2) alone is not sufficient to answer the question asked.
(B) Statement (2) ALONE is sufficient, but statement (1) alone is not sufficient to answer the question asked.
(C) BOTH statements (1) and (2) TOGETHER are sufficient to answer the question asked; but NEITHER statement ALONE is sufficient.
(D) EACH statement ALONE is sufficient to answer the question asked.
(E) Statements (1) and (2) TOGETHER are NOT sufficient to answer the question asked, and additional data specific to the problem are needed.

3. Kevin buys beer in bottles and cans. He pays \$1.00 for each can of beer and \$1.50 for each bottle of beer. If he buys a total of 15 bottles and cans of beer, how many bottles of beer did he buy?

(1) Kevin spent a total of \$18.00 on beer.

(2) Kevin bought 3 more cans of beer than bottles of beer.

(A) Statement (1) ALONE is sufficient, but statement (2) alone is not sufficient to answer the question asked.
(B) Statement (2) ALONE is sufficient, but statement (1) alone is not sufficient to answer the question asked.
(C) BOTH statements (1) and (2) TOGETHER are sufficient to answer the question asked; but NEITHER statement ALONE is sufficient.
(D) EACH statement ALONE is sufficient to answer the question asked.
(E) Statements (1) and (2) TOGETHER are NOT sufficient to answer the question asked, and additional data specific to the problem are needed.

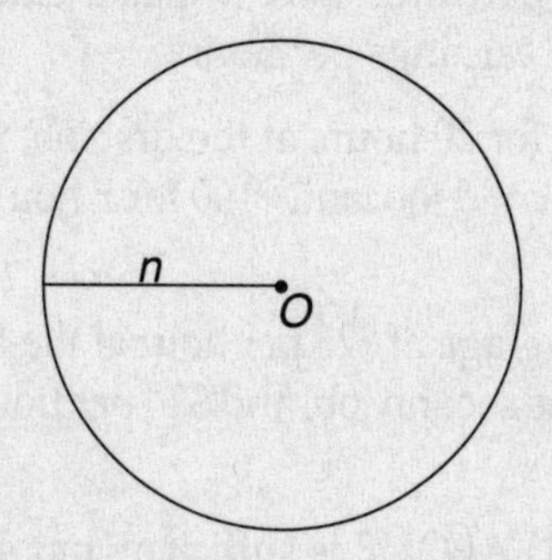


4. What is the area of the circle with center $O$ above?

   (1) The circumference of the circle is $12\pi$.

   (2) $n = 6$

   (A) Statement (1) ALONE is sufficient, but statement (2) alone is not sufficient to answer the question asked.
   (B) Statement (2) ALONE is sufficient, but statement (1) alone is not sufficient to answer the question asked.
   (C) BOTH statements (1) and (2) TOGETHER are sufficient to answer the question asked; but NEITHER statement ALONE is sufficient.
   (D) EACH statement ALONE is sufficient to answer the question asked.
   (E) Statements (1) and (2) TOGETHER are NOT sufficient to answer the question asked, and additional data specific to the problem are needed.

5. Pete works at three part-time jobs to make extra money. What are his average earnings per hour?

   (1) Pete earned \$500 for 20 hours at the first job, \$150 for 10 hours at the second job, and \$100 for 5 hours at the third job.

   (2) Pete earned an average of \$25 per hour at the first job, \$15 per hour at the second job, and \$20 per hour at the third job.

   (A) Statement (1) ALONE is sufficient, but statement (2) alone is not sufficient to answer the question asked.
   (B) Statement (2) ALONE is sufficient, but statement (1) alone is not sufficient to answer the question asked.
   (C) BOTH statements (1) and (2) TOGETHER are sufficient to answer the question asked; but NEITHER statement ALONE is sufficient.
   (D) EACH statement ALONE is sufficient to answer the question asked.
   (E) Statements (1) and (2) TOGETHER are NOT sufficient to answer the question asked, and additional data specific to the problem are needed.

6. If $p$ is an integer, is $p$ positive?

   (1) $pq > 0$ and $qr < 0$

   (2) $pr < 0$

   (A) Statement (1) ALONE is sufficient, but statement (2) alone is not sufficient to answer the question asked.
   (B) Statement (2) ALONE is sufficient, but statement (1) alone is not sufficient to answer the question asked.
   (C) BOTH statements (1) and (2) TOGETHER are sufficient to answer the question asked; but NEITHER statement ALONE is sufficient.
   (D) EACH statement ALONE is sufficient to answer the question asked.
   (E) Statements (1) and (2) TOGETHER are NOT sufficient to answer the question asked, and additional data specific to the problem are needed.

7. If $x$ is an integer, what is the value of $x$ ?

(1) $x$ is the square of an integer.

(2) $0 < x < 5$

(A) Statement (1) ALONE is sufficient, but statement (2) alone is not sufficient to answer the question asked.
(B) Statement (2) ALONE is sufficient, but statement (1) alone is not sufficient to answer the question asked.
(C) BOTH statements (1) and (2) TOGETHER are sufficient to answer the question asked; but NEITHER statement ALONE is sufficient.
(D) EACH statement ALONE is sufficient to answer the question asked.
(E) Statements (1) and (2) TOGETHER are NOT sufficient to answer the question asked, and additional data specific to the problem are needed.

8. If $m$ and $n$ are integers, is $mn \leq 6$?

(1) $m + n = 5$

(2) $1 \leq m \leq 3$ and $2 \leq n \leq 4$

(A) Statement (1) ALONE is sufficient, but statement (2) alone is not sufficient to answer the question asked.
(B) Statement (2) ALONE is sufficient, but statement (1) alone is not sufficient to answer the question asked.
(C) BOTH statements (1) and (2) TOGETHER are sufficient to answer the question asked; but NEITHER statement ALONE is sufficient.
(D) EACH statement ALONE is sufficient to answer the question asked.
(E) Statements (1) and (2) TOGETHER are NOT sufficient to answer the question asked, and additional data specific to the problem are needed.

9. If $2x + 3y = 11$, what is the value of $x$ ?

(1) $5x - 2y = 18$

(2) $6y - 22 = -4x$

(A) Statement (1) ALONE is sufficient, but statement (2) alone is not sufficient to answer the question asked.
(B) Statement (2) ALONE is sufficient, but statement (1) alone is not sufficient to answer the question asked.
(C) BOTH statements (1) and (2) TOGETHER are sufficient to answer the question asked; but NEITHER statement ALONE is sufficient.
(D) EACH statement ALONE is sufficient to answer the question asked.
(E) Statements (1) and (2) TOGETHER are NOT sufficient to answer the question asked, and additional data specific to the problem are needed.

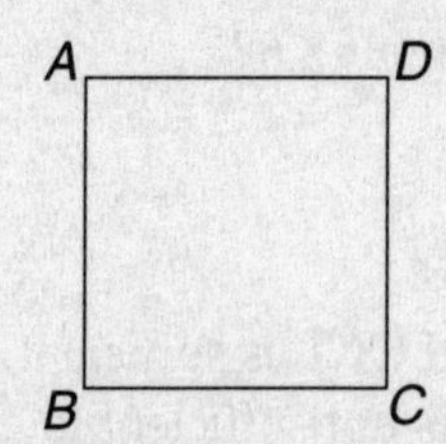


10. Is quadrilateral $ABCD$ above a square?

(1) $AB = BC$

(2) Angle $ABC$ is a right angle.

(A) Statement (1) ALONE is sufficient, but statement (2) alone is not sufficient to answer the question asked.
(B) Statement (2) ALONE is sufficient, but statement (1) alone is not sufficient to answer the question asked.
(C) BOTH statements (1) and (2) TOGETHER are sufficient to answer the question asked; but NEITHER statement ALONE is sufficient.
(D) EACH statement ALONE is sufficient to answer the question asked.
(E) Statements (1) and (2) TOGETHER are NOT sufficient to answer the question asked, and additional data specific to the problem are needed.

## PRACTICE SET — EXPLANATIONS AND ANSWERS

1. (C) is correct. Look at statement (1). If you factor it, you get $(n + 2)(n + 3) = 0$ which means that $n = -2$ or $-3$. That's not a single value, so you can't answer the question. Narrow the choices to (B), (C), or (E). Try statement (2). This factors to $(n + 2)(n - 3) = 0$, so $n = -2$ or 3. This isn't a single value, so you can't answer the question. Eliminate (B). Try (1) and (2) together. The only number that is on both lists is $-2$, so $n = -2$. You can answer the question, so choose (C).

2. (C) is correct. Look at statement (1). If you plug in a positive number for $y$, such as $y = 2$, then $x = y$ and the answer is "yes." Same thing if you use a negative number, such as $y = -2$. However, if plug in $y = 0$, then $x$ could equal anything, so "no" is a possibility. You can't answer the question, so narrow the choices to (B), (C), or (E). Try statement (2). If you plug in two positive numbers, such as $x = 2$ and $y = 2$, then $x = y$ and the answer is "yes." If you plug in a positive number and a negative number, such as $x = 2$ and $y = -2$, then the answer is "no." You can't answer the question, so eliminate (B). Try (1) and (2) together. The only numbers that will fit both statements are those in which $x = y$. You can't use a positive and a negative because that won't fit statement (1). You can't use 0 for $y$ and something else for $x$, because that won't fit statement (2). It must be true that $x = y$. The answer is "yes," so choose (C).

3. (D) is correct. The question provides two variables, bottles and cans, and one equation, bottles + cans = 15. Look at statement (1). This provides another equation, $(1 \times \text{cans}) + (1.5 \times \text{bottles}) = 18$. You have two equations for two variables, so you can answer the question. Narrow the choices to (A) or (D). Look at statement (2). This provides another equation, cans – 3 = bottles. You have two equations for two variables, so you can answer the question. Choose (D).

4. (D) is correct. You know that the formula for the area of a circle is $A = \pi r^2$, so the missing piece is the radius. Look for that as you go through the statements. In statement (1), you're given the circumference, so you can solve for the radius by plugging $C = 12\pi$ into the equation $C = 2\pi r$. You can answer the question, so narrow the choices to (A) or (D). Look at statement (2). This gives you the radius. You can answer the question, so choose (D).

5. (A) is correct. To find an average, you need to know the total and the number of elements. So the missing pieces are the total money earned and the number of hours. Look for those in the statements. In statement (1) you can find both the total money earned and the number of hours worked. You can answer the question, so narrow the choices to (A) or (D). Look at statement (2). You can find neither the total money earned nor the number of hours worked. Remember, never average the averages. You can't answer the question, so choose (A).

6. (E) is correct. Look at statement (1). You could plug in positive numbers for $p$ and $q$ and that would make $r$ negative. The answer would be "yes." Or you could plug in negative numbers for $p$ and $q$ and that would make $r$ positive. In that case, the answer would be "no." You can't answer the question, so narrow the choices to (B), (C), or (E). Look at statement (2). You could plug in a positive number for $p$ and a negative number for $r$ and the answer would be "yes." Or you could plug in a negative number for $p$ and a positive number for $r$ and the answer would be "no." You can't answer the question, so eliminate (B). Try (1) and (2) together. You could plug in $p = +$, $q = +$, and $r = -$. That would fit both statements and the answer would be "yes." Or you could plug in $p = -$, $q = -$, and $r = +$. That would fit both statements and the answer would be "no." You can't answer the question, so choose (E).

7. (E) is correct. Look at statement (1). $x$ must be a perfect square, such as 1, 4, 9, 16, etc. However, you can't narrow it down to a single value. You can't answer the question, so the possible choices are (B), (C), or (E). Look at statement (2). $x$ could be 1, 2, 3, or 4. You can't answer the question, so eliminate (B). Try (1) and (2) together. Although the lists overlap, they share more than one value. $x$ could be either 1 or 4. You can't answer the question, so choose (E).

8. (A) is correct. Look at statement (1). If you plug in $m = 1$ and $n = 4$, then $mn = 4$ and the answer is "yes." If you plug in $m = 2$ and $n = 3$, then $mn = 6$ and the answer is "yes." If you plug in $m = -5$ and $n = 10$, then $mn = -50$ and the answer is "yes." Any numbers that will fit the statement will give you a "yes" answer. You can answer the question, so narrow the choices to (A) or (D). Look at statement (2). If you plug in $m = 1$ and $n = 2$, then $mn = 2$ and the answer is "yes." If you plug in $m = 3$ and $n = 4$, then $mn = 12$ and the answer is "no." You can't answer the question, so choose (A).

9. (A) is correct. The question provides one equation and two variables. Look at statement (1). This gives you another equation. With two equations for two variables, you can solve for $x$. You can answer the question, so narrow the choices to (A) or (D). Look at statement (2). Although this seems to give you another equation, it really doesn't. The equation in (2) is the equation from the question rearranged and multiplied by 2. If you tried to set up the simultaneous equations and solve them, everything would cancel and you'd be stuck. Remember, you need a distinct equation for each variable. You can't answer the question so choose (A).

10. (E) is correct. Look at statement (1). It's possible that ABCD is a square. You could make all four sides equal and make all four angles right angles. However, it's also possible that it's not a square. Just because $AB = BC$ doesn't mean that the other two sides are also the same. Don't let the diagram fool you. It's not necessarily drawn to scale. You can't answer the question, so narrow the choices to (B), (C), or (E). Look at statement (2). By itself, this doesn't tell you much. It might be a square, but you don't know whether the sides are all the same or whether the other angles are also right angles. You can't answer the question, so eliminate (B). Try (1) and (2) together. It might be a square, but it doesn't have to be. The other three angles don't have to be right angles and *CD* and *AD* might not equal *AB* and *BC*. See the diagram below. You can't answer the question, so choose (E).

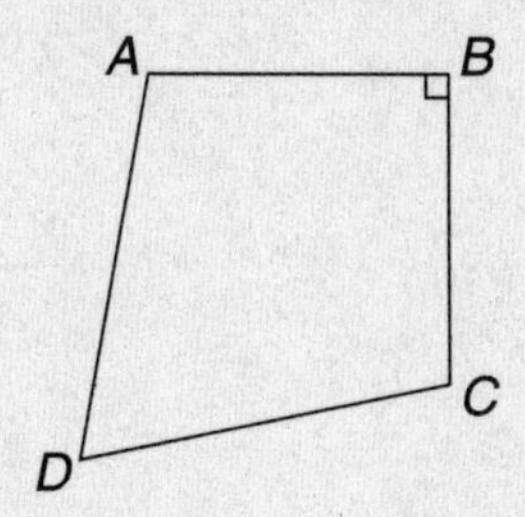

# 11

# Assorted Other Topics #2

## PROBABILITY

The probability that a particular event will occur is usually expressed as a fraction. For example, the probability that a flipped coin will land heads up is $\frac{1}{2}$. To calculate the probability that several things will occur, just multiply the fractions of each individual probability. For example, the probability of getting heads on three coin flips is $\frac{1}{2} \times \frac{1}{2} \times \frac{1}{2} = \frac{1}{8}$. This assumes that the events are independent. In other words, the result of the first coin flip doesn't affect the other coin flips. This will be true on the probability questions on the GMAT. Also, it doesn't matter whether you flip one coin three times or flip three coins once. The probability is the same in both cases. Try this next example:

1. Two dice, each with six sides numbered 1 through 6, are tossed. What is the probability that both dice will come up with either a 1 or a 2?

   (A) $\frac{1}{36}$

   (B) $\frac{1}{18}$

   (C) $\frac{1}{9}$

   (D) $\frac{1}{3}$

   (E) $\frac{2}{3}$

The probability that the first die will result in a 1 or a 2 is 2 out of 6, or $\frac{2}{6} = \frac{1}{3}$. The probability is the same for the second die. To get the probability that both events will occur, just multiply the two probabilities: $\frac{1}{3} \times \frac{1}{3} = \frac{1}{9}$. So the answer is (C).

In some cases, the probability of later events changes because things get "used up." Suppose there are 4 marbles in a jar, 3 blue and 1 red. What is the probability of drawing 2 blue marbles from the jar? Well the probability that the first marble is blue is $\frac{3}{4}$. However, there are now only 3 marbles left, 2 blue and 1 red (assume that you successfully picked a blue marble the first time). So the probability that the second marble is blue is $\frac{2}{3}$.

So the probability of picking 2 blue marbles is $\frac{3}{4} \times \frac{2}{3} = \frac{1}{2}$. Try this next example:

2. Frank has a box containing 6 doughnuts: 3 jellies and 3 cinnamon twists. If Frank eats three doughnuts, chosen randomly from the box, what is the probability that he eats three jelly doughnuts?

(A) $\frac{1}{2}$

(B) $\frac{1}{3}$

(C) $\frac{1}{8}$

(D) $\frac{1}{20}$

(E) $\frac{1}{36}$

The probability that the first doughnut is a jelly is $\frac{1}{2}$. Assuming he gets a jelly, the probability that the second doughnut is a jelly is $\frac{2}{5}$, because there 2 jellies left out of 5 doughnuts. Assuming he got 2 jellies already, the probability that the third is a jelly is $\frac{1}{4}$ because there is 1 jelly left out of 4 doughnuts. To find the probability that all three events happened, multiply the probabilities: $\frac{1}{2} \times \frac{2}{5} \times \frac{1}{4} = \frac{1}{20}$. The answer is (D).

Sometimes it's easier to calculate the probability that something will *not* happen than to calculate the probability that it will happen. If the probability that something *will happen* is $x$, then the probability that it *won't happen* is $1 - x$. For example, suppose there is a 40% chance that it will rain tomorrow. That means there is a $1 - 0.4 = 0.6$ or 60% chance that it will not rain tomorrow. Suppose you need to find the probability that you will get at least 1 heads when you flip a coin 5 times. The hard way to do it would be to find the probabilities of getting exactly 1 heads, exactly 2 heads, exactly 3 heads, and exactly 4 heads and then add them all up. The easy way is to find the probability that you do *not* get at least 1 heads. That means you get 5 tails. The probability of that is simply $\frac{1}{2} \times \frac{1}{2} \times \frac{1}{2} \times \frac{1}{2} \times \frac{1}{2} = \frac{1}{32}$. So the probability that you will get at least 1 heads is $1 - \frac{1}{32} = \frac{31}{32}$.

Try this next example:

3. A six-sided die, with sides numbered 1 through 6, is rolled three times. What is the probability that the sum of the three rolls is at least 4?

(A) $\frac{1}{216}$

(B) $\frac{1}{54}$

(C) $\frac{1}{2}$

(D) $\frac{53}{54}$

(E) $\frac{215}{216}$

This problem is easier if you start with the probability that the sum of the rolls is less than 4. The only way that can happen is to get a 1 on each of the three rolls. The probability of getting a 1 is $\frac{1}{6}$, so the probability of getting a 1 three times is $\frac{1}{6}\times\frac{1}{6}\times\frac{1}{6}=\frac{1}{216}$. The probability that you will get at least 4 is $1-\frac{1}{216}=\frac{215}{216}$. The correct answer is (E).

## FUNCTIONS

These problems usually contain some bizarre symbols or strange terms. A question might test a function such as $x \# y = x^2 + y^2 + 2xy$ or use an odd term such as *hyper-prime*. However, a function problem is really just an exercise in following directions. The question must define the function for you, either as a formula or as a description of how to manipulate the numbers. Just plug the numbers they give you into that definition. Here's an example:

1. If $a$ @ $b = 3a^2 + 2b - 1$, then 2 @ 3 =

(A) 12
(B) 17
(C) 30
(D) 37
(E) 41

Just plug in $a = 2$ and $b = 3$ into the formula the question provides. You get 2 @ 3 = $3(2)^2 + 2(3) - 1 = 12 + 6 - 1 = 17$. The answer must be (B).

Sometimes the GMAT will use strange term to indicate a function, rather than a symbol. You may even see something that has a real-life use, such as the net present value of an annuity or some other financial function. Don't worry. The question will provide the directions, either as a formula or a step-by-step description. All you have to do is plug in the numbers and do the calculation.

## RATES

A number of GMAT questions deal with how fast someone can complete a task (e.g. paint a wall or build widgets) or how fast someone travels. These problems all revolve around that person's rate. The formula for rate can be expressed a couple of ways:

$$\text{Amount} = \text{Rate} \times \text{Time}$$

$$\text{Rate} = \frac{\text{Amount}}{\text{Time}}$$

Just fill in the numbers the question provides and solve for the other one. Look at this example:

1. If Albert can travel 200 miles in 4 hours, how many hours will it take Albert, traveling at the same constant rate, to travel 350 miles?

   (A) 5
   (B) 6
   (C) 7
   (D) 8
   (E) 10

The key number is usually the rate. Find it if the question doesn't state it. Albert's rate (or speed) is $\frac{200}{4} = 50$ per hour. He needs to travel 350 miles. Plug those numbers into the formula to get $350 = 50 \times \text{Time}$, which is solved as $\text{Time} = \frac{350}{50} = 7$ hours. The answer is (C).

If several people are working together, just add their individual rates to find their combined rate. For example, if Joe can paint 10 square feet per hour and Jack can paint 8 square feet per hour, then they paint 18 square feet per hour when they work together. This also applies if two people are traveling toward each other. Add their rates to find how quickly they close the gap. Suppose Alice and Brenda are 300 miles apart and begin driving toward each other. If Alice drives at a speed of 40 miles per hour and Brenda drives at a speed of 60 miles per hour, they get closer at a rate of 40 + 60 = 100 miles per hour. So they will meet in $\frac{300}{100} = 3$ hours. Try this example:

2. Mark can process 30 insurance claims per hour. Bruce can process 15 claims per hour. Mark starts working on a batch of 555 insurance claims. Two hours later, Bruce begins working with Mark until the batch is finished. How many hours did Mark spend working on the batch?

   (A) 2
   (B) 11
   (C) 13
   (D) 22
   (E) 26

If Mark works alone for 2 hours, he processes $2 \times 30 = 60$ claims, leaving $555 - 60 = 495$ claims to go. When Bruce joins Mark, they work at the combined rate of 30 + 15 = 45 claims per hour, so they need $\frac{495}{45} = 11$ hours to finish the batch. So Mark spends a total of 2 (alone) + 11 (with Bruce) = 13 hours working on the batch. The correct answer is (C).

If the task or the distance is undefined, you may want to use the Plugging In technique. Make up a number for the amount of work or the distance and solve the problem from there. Try this example:

3. Working alone, Bud can complete a particular task in 6 hours. Lou, working alone, can complete the same task in 8 hours. If Bud and Lou work together, how many hours will it take them to complete the task?

   (A) 7

   (B) $3\frac{4}{7}$

   (C) $3\frac{1}{2}$

   (D) $3\frac{3}{7}$

   (E) 3

The task is undefined, so make up a number for it. Suppose the task is 48 units (because it works well with 6 and 8). Bud can complete the task in 6 hours, so his individual rate is $\frac{48}{6} = 8$ units per hour. Lou's rate is $\frac{48}{8} = 6$ units per hour. Combined, they work at a rate of 8 + 6 = 14 units per hour. To complete 48 units, they will need $\frac{48}{14} = \frac{24}{7} = 3\frac{3}{7}$ hours. The correct answer is (D). Note that the answer is not $\frac{6+8}{2} = 7$ hours or $\frac{1}{2} \times 7 = 3\frac{1}{2}$ hours. Don't fall for those Joe Bloggs traps. Use the formula and work it out.

## DIGIT PROBLEMS

In these problems, one digit of a number is replaced by a symbol or letter. For example, the problem may contain a number such as 11,8#3, where # represents the missing digit. Usually, you need to find the value or the range of possible values for the missing digit. Remember that there are only 10 possible values for a single digit: 0 through 9.

You can usually solve a digit problem by applying the Backsolving technique. Try the numbers from the answer choices until you find one that fits. Look at this example:

```
  510
×   #
-----
2,##0
```

1. In the multiplication above, # represents a single digit. What digit does # represent?

   (A) 2
   (B) 3
   (C) 4
   (D) 5
   (E) 6

Just try out the different answers until you find one that fits. If you plug in # = 2, the product is 1,020, which doesn't fit. With # = 3, the product is 1,420. That doesn't fit either. Plug in # = 4 and the product is 2,040. That doesn't work. With # = 5, the product is 2,550. That works, so the answer is (D).

## COMBINATIONS AND PERMUTATIONS

These problems deal with the number of different ways to arrange a group of things or the number of different selections you can make from a group of things. One critical distinction to make is between problems in which the order of the things *does* matter and problems in which the order of the things *does not* matter.

Suppose you're scheduling speakers for a seminar. You have three people—A, B, and C—available but you only need two different people, one for the morning and one for the afternoon. Your choices are A then B, A then C, B then A, B then C, C then A, and C then B. That's a total of six different pairs. Notice that order does matter. You want to count "A then B" and "B then A" as two different pairs.

Now suppose that you're choosing fruits to make a health shake. You need two different fruits and you have three from which to choose—A, B, and C. You could choose A and B, A and C, or B and C. That's only three different pairs (not pears). You don't count "A and B" and "B and A" as two different pairs because they're just going into the blender together anyway.

When order matters, the different arrangements are called **permutations**. When order doesn't matter, the different arrangements are called **combinations**. The names aren't that important, but you do need to understand the difference and be able to identify which one the question wants.

In the examples above, you could just list out the permutations and combinations. However, sometimes there are too many for that be an effective strategy. There are formulas for calculating the number of permutations and combinations. The formula for the number of permutations is:

$$\frac{n!}{(n-r)!}$$

Permutations (Order does matter)

The $n$ is the number of things you have to choose from (3 in the examples above) and $r$ is the number of spots you need to fill (2 in the examples above). The ! is the symbol for a **factorial**. To calculate a factorial, just multiply that number by every lower number, all the way down to 1. For example, $3! = 3 \times 2 \times 1$ and $2! = 2 \times 1$. Just list out the factorials and cancel the numbers. Then reduce the fraction to get the number of permutations. For the seminar speakers example, the formula would be $\frac{3!}{(3-2)!} = \frac{3!}{1!} = \frac{3 \times 2 \times 1}{1} = 6$.

The formula for combinations is very similar. It uses the same $n$ and $r$ and also requires factorials. Here's the formula:

$$\frac{n!}{r!\,(n-r)!}$$

Combinations (Order does not matter)

The calculation for the fruit shake example would be $\frac{3!}{2! \times (3-2)!} = \frac{3!}{2! \times 1!} = \frac{3 \times 2 \times 1}{2 \times 1 \times 1} = \frac{6}{2} = 3$.

These formulas take a little practice, but they're not quite rocket science. They are good things to "brain dump" onto your scratch paper at the beginning of the math section. Try this example for practice:

1. Eric is ordering a pizza. He wants three different toppings on his pizza. The available toppings are pepperoni, sausage, black olives, green peppers, and mushrooms. How many different combinations of toppings are possible?

   (A) 5
   (B) 10
   (C) 15
   (D) 30
   (E) 60

In this example, order does not matter. Eric only cares which three toppings are chosen, not the order they are chosen. So you should use the combination formula. He is choosing 3 items from a group of 5. Plugging the numbers into the formula, you get

$\frac{5!}{3!\times(5-3)!}=\frac{5!}{3!\times 2!}=\frac{5\times 4\times 3\times 2\times 1}{3\times 2\times 1\times 2\times 1}=10$. So the answer is (B).

The formulas for permutations and combinations both assume that you can only use each thing in the group once. If you're allowed to reuse things, you'll have to skip the formulas and rely on a more general approach. The following example will illustrate the general approach:

2. Mike needs to choose a computer password. The password will consist of two letters followed by three numerical digits. The letters may be the same, but no digit may match another digit. How many different passwords could Mike create?

To solve this problem, write out five blanks, one for each letter or digit that Mike needs to choose. In each blank, write the number of choices he has for that blank. Keep in mind the restrictions. Then multiply all the numbers. This gives you the equivalent of a permutation (order does matter).

For the first blank, Mike has 26 letters (the alphabet) from which to choose. For the second blank, he also has 26 choices because he is allowed to reuse letters. For the third blank, he has 10 digits (0 through 9) from which to choose. For the fourth blank, he has only 9 digits because he can't use the digit from the third blank. For the fifth blank, he has only 8 digits left. So Mike has $26\times 26\times 10\times 9\times 8=486{,}720$ possible passwords.

## MEDIAN, MODE, AND RANGE

Back in Chapter 6, you learned how to find the average of a group of numbers. The GMAT also tests some close cousins of the average, namely the median, mode, and range of a set of numbers.

To calculate the **median** of a group of numbers, first arrange the numbers in order, either ascending or descending. The median is simply the middle number. If there is an even number of things in the group, there won't be a single middle number. In that case, the median is the average of the two middlemost numbers.

The **mode** of a group of numbers is the number that appears the most number of times. If every number is different, there is no mode.

The **range** of a set of numbers is the difference between the greatest number in the set and the least number in the set. Just subtract.

Suppose you have this set of numbers: 1, 5, –5, 3, 1, 2, and 8. To find the median, arrange them in order: –5, 1, 1, 2, 3, 5, and 8. The number in the middle is 2, so the median is 2. The mode in this set is 1 because it shows up twice while each of the other numbers show up once. The range of the set is 8 – (–5) = 13.

## PRACTICE SET

1. Alex drives to and from work each day along the same route. If he drives at a speed of 80 miles per hour on the way to work and he drives at a speed of 100 miles per hour on the way from work, which of the following most closely approximates his average speed in miles per hour for the round trip?

   (A) 80.0
   (B) 88.9
   (C) 90.0
   (D) 91.1
   (E) 100.0

2. When a certain coin is flipped, the probability that it will land on heads is $\frac{1}{2}$ and the probability that it will land on tails is $\frac{1}{2}$. If the coin is flipped three times, what is the probability that all three results are the same?

   (A) $\frac{1}{8}$
   (B) $\frac{1}{4}$
   (C) $\frac{3}{8}$
   (D) $\frac{1}{2}$
   (E) $\frac{7}{8}$

3. Joe is choosing books at the bookstore. He has a list of seven books that he would like to buy, but he can only afford to buy three books. How many different groups of books could Joe buy?

   (A) 210
   (B) 155
   (C) 70
   (D) 35
   (E) 21

$$\begin{array}{r} \# \\ + \; @ \\ \hline \& \end{array}$$

4. In the addition above, #, @, and & each represent a distinct, positive digit. If & is even, what is the value of #?

   (1) @ is even and & = 6

   (2) # < @

   (A) Statement (1) ALONE is sufficient, but statement (2) alone is not sufficient to answer the question asked.
   (B) Statement (2) ALONE is sufficient, but statement (1) alone is not sufficient to answer the question asked.
   (C) BOTH statements (1) and (2) TOGETHER are sufficient to answer the question asked; but NEITHER statement ALONE is sufficient.
   (D) EACH statement ALONE is sufficient to answer the question asked.
   (E) Statements (1) and (2) TOGETHER are NOT sufficient to answer the question asked, and additional data specific to the problem are needed.

5. If Max can complete a job in 4 hours and Nick can complete the same job in 6 hours, how many fewer hours do Max and Nick working together need to complete the job than Max alone needs to complete the job?

   (A) 1.6
   (B) 2.4
   (C) 3.2
   (D) 3.4
   (E) 5.0

6. If a set of numbers consists of 10, 15, 0 , 3, and $x$, and the range of the set is 30, what are the possible values for the median of the set?

   (A) –15 and 30
   (B) 15 and 10
   (C) 0 and 3
   (D) 3 and 15
   (E) 3 and 10

7. If the area of a number is defined as the difference between that number's greatest and least prime factors, what is the area of 100?

(A) 0
(B) 2
(C) 3
(D) 5
(E) 9

8. Carol is having friends over to watch home movies. She has 6 reels of home movies, but she and her friends have time to watch only 3 of them. Carol must decide which reels to watch and in what order. How many different orderings does Carol have from which to choose?

(A) 120
(B) 96
(C) 60
(D) 36
(E) 20

9. A jar contains 5 marbles: 3 red and 2 blue. If two marbles are drawn randomly from the jar, what is the probability that they will be different colors?

(A) $\frac{3}{25}$

(B) $\frac{6}{25}$

(C) $\frac{2}{5}$

(D) $\frac{3}{10}$

(E) $\frac{3}{5}$

10. Oscar is running in a straight line away from Nancy at the rate of 20 feet per second. Nancy is chasing Oscar at the rate of 25 feet per second. If Oscar has a 100-foot head start, how long, in seconds, will it take Nancy to catch Oscar?

(A) 4
(B) 5
(C) 10
(D) 20
(E) 100

## PRACTICE SET — ANSWERS AND EXPLANATIONS

1. (B) is correct. The distance is undefined, so make one up. Suppose his trip is 400 miles each way. On the way to work, he travels at 80 miles per hour and completes the trip in $\frac{400}{80} = 5$ hours. On the way from work, he travels 100 miles per hour and completes the trip in $\frac{400}{100} = 4$ hours. For the round trip, he travels 400 + 400 = 800 miles in 5 + 4 hours. That's an average speed of $\frac{800}{9} = 88.88$ miles per hour. (B) is the correct answer.

2. (B) is correct. It doesn't really matter which way the first flips ends up, so don't count that probability. The probability that the second flip is the same as the first is $\frac{1}{2}$. The probability that the third flip matches is also $\frac{1}{2}$. So the probability that the last two flips match the first is $\frac{1}{2} \times \frac{1}{2} = \frac{1}{4}$. Choose (B).

3. (D) is correct. Order does not matter, so use the combination formula. Plug In 7 for $n$ and 3 for $r$ to get $\frac{7!}{3! \times (7-3)!} = \frac{7!}{3! \times 4!} = \frac{7 \times 6 \times 5 \times 4 \times 3 \times 2 \times 1}{3 \times 2 \times 1 \times 4 \times 3 \times 2 \times 1} = 35$ combinations. Choose (D).

4. (C) is correct. Look at statement (1). If # and @ are both even, positive digits that add up to 6, then one of them is 2 and one of them is 4. However, you can't tell which one is which. You can't answer the question, so narrow the choices to (B), (C), or (E). Look at statement (2). This isn't much help by itself. There are several possible values for #, @, and &. You can't answer the question, so eliminate (B). Try (1) and (2) together. From (1), you know that # and @ are 2 and 4. From (2), you know that # is the smaller of the two. So # = 2 and @ = 4. You can answer the question, so choose (C).

5. (A) is correct. The task is undefined, so make up a value, say 48 units. Max can complete 48 units in 4 hours so his rate is $\frac{48}{4} = 12$ units per hour. Nick can complete 48 units in 6 hours, so his rate is $\frac{48}{6} = 8$ units per hour. Combined, Max and Nick work at the rate of 12 + 8 = 20 units per hour and can complete the job in $\frac{48}{20} = 2.4$ hours. That's 4 – 2.4 = 1.6 fewer hours than Max alone needs. Choose (A).

6. (E) is correct. For the range to be 30, either $x$ is the highest number and $x = 30$ or $x$ is the lowest number and $x = -15$. If $x = 30$, the set is 0, 3, 10, 15, and 30, so the median would be 10. If $x = -15$, the set is –15, 0, 3, 10, 15, so the median would be 3. So the possible values for the median are 3 and 10. Choose (E).

7. (C) is correct. The prime factorization of 100 is $2 \times 2 \times 5 \times 5$. So the greatest prime factor is 5 and the least is 2. That's a difference of $5 - 2 = 3$. Choose (C).

8. (A) is correct. Order is important, so use the permutation formula. Plug in $n = 6$ and $r = 3$ to get $\frac{6!}{(6-3)!} = \frac{6!}{3!} = \frac{6 \times 5 \times 4 \times 3 \times 2 \times 1}{3 \times 2 \times 1} = 120$. Choose (A).

9. (E) is correct. The probability that the first marble will be red is $\frac{3}{5}$. Assuming the first one is red, the probability that the second marble will be blue is $\frac{1}{2}$ because there are 4 marbles left, 2 of which are blue. So the probability of getting red then blue is $\frac{3}{5} \times \frac{1}{2} = \frac{3}{10}$. However, getting blue then red would also be acceptable. The probability that the first marble is blue is $\frac{2}{5}$. Assuming that the first marble is blue, the probability that the second marble is red is $\frac{3}{4}$, because 3 of the remaining 4 marbles are red. So the probability of getting blue then red is $\frac{2}{5} \times \frac{3}{4} = \frac{3}{10}$. Because either of these patterns is acceptable, add the probabilities to find the probability that the two marbles are different colors: $\frac{3}{10} + \frac{3}{10} = \frac{6}{10} = \frac{3}{5}$. Choose (E).

10. (D) is correct. If Oscar is traveling 20 feet per second and Nancy is traveling 25 feet per second, then Nancy is gaining on Oscar at the rate of $25 - 20 = 5$ feet per second. So it will take her $\frac{100}{5} = 20$ seconds to close the 100-foot gap. Choose (D).

# FIND US...

## International

**Hong Kong**
4/F Sun Hung Kai Centre
30 Harbour Road, Wan Chai,
Hong Kong
Tel: (011)85-2-517-3016

**Japan**
Fuji Building 40, 15-14
Sakuragaokacho, Shibuya Ku,
Tokyo 150, Japan
Tel: (011)81-3-3463-1343

**Korea**
Tae Young Bldg, 944-24,
Daechi- Dong, Kangnam-Ku
The Princeton Review- ANC
Seoul, Korea 135-280,
South Korea
Tel: (011)82-2-554-7763

**Mexico City**
PR Mex S De RL De Cv
Guanajuato 228 Col. Roma
06700 Mexico D.F., Mexico
Tel: 525-564-9468

**Montreal**
666 Sherbrooke St.
West, Suite 202
Montreal, QC H3A 1E7 Canada
Tel: (514) 499-0870

**Pakistan**
1 Bawa Park - 90 Upper Mall
Lahore, Pakistan
Tel: (011)92-42-571-2315

**Spain**
Pza. Castilla, 3 - 5° A, 28046
Madrid, Spain
Tel: (011)341-323-4212

**Taiwan**
155 Chung Hsiao East Road
Section 4 - 4th Floor,
Taipei R.O.C., Taiwan
Tel: (011)886-2-751-1243

**Thailand**
Building One, 99 Wireless Road
Bangkok, Thailand 10330
Tel: (662) 256-7080

**Toronto**
1240 Bay Street, Suite 300
Toronto M5R 2A7 Canada
Tel: (800) 495-7737
Tel: (716) 839-4391

locations

**Vancouver**
4212 University Way NE,
Suite 204
Seattle, WA 98105
Tel: (206) 548-1100

## National (U.S.)

We have over 60 offices around the U.S. and run courses in over 400 sites. For courses and locations within the U.S. call 1 (800) 2/Review and you will be routed to the nearest office.